D0960223

ALSO BY JONATHAN SAFRAN FOER

FICTION

Here I Am

Extremely Loud and Incredibly Close

Everything Is Illuminated

NONFICTION

Eating Animals

OTHER

Tree of Codes

Joe (with Hiroshi Sugimoto)

The Long Never (with Hiroshi Sugimoto)

Seven Attempted Escapes from Silence: A Libretto

AS EDITOR

A Convergence of Birds

New American Haggadah

WE ARE
THE WEATHER

WE ARE THE WEATHER

Saving the Planet Begins at Breakfast

JONATHAN SAFRAN FOER

Farrar, Straus and Giroux
New York

Farrar, Straus and Giroux
120 Broadway, New York 10271

Printed in the United States of America
First edition, 2019

Library of Congress Cataloging-in-Publication Data
Names: Foer, Jonathan Safran, 1977– author.
Title: We are the weather : saving the planet begins at breakfast / Jonathan Safran Foer.
Description: First edition. | New York : Farrar, Straus and Giroux, 2019. | Includes bibliographical references.
Identifiers: LCCN 2019020218 | ISBN 9780374280000 (hardcover)
Subjects: LCSH: Livestock—Climatic factors. | Animal culture. | Sustainable agriculture.
Classification: LCC SF140.C57 F64 2019 | DDC 636—dc23
LC record available at https://lccn.loc.gov/2019020218

Open Market edition ISBN: 978-0-374-90954-3

Designed by Jonathan D. Lippincott

www.fsgbooks.com
www.twitter.com/fsgbooks • www.facebook.com/fsgbooks

1 3 5 7 9 10 8 6 4 2

For Sasha and Cy, Sadie and Theo, Leo and Bea

Contents

I. UNBELIEVABLE

The Book of Endings

The oldest suicide note was written in ancient Egypt about four thousand years ago. Its original translator titled it "Dispute with the Soul of One Who Is Tired of Life." The first line reads, "I opened my mouth to my soul, that I might answer what it said." Careening between prose, dialogue, and poetry, what follows is a person's effort to persuade his soul to consent to suicide.

I learned about that note from *The Book of Endings*, a compilation of facts and anecdotes that also includes the dying wishes of Virgil and Houdini; elegies to the dodo and the eunuch; and explanations of the fossil record, the electric chair, and man-made obsolescence. I wasn't a particularly morbid child, but for years I carried that morbid paperback around with me.

The Book of Endings also taught me that my every inhalation includes molecules from Julius Caesar's final exhalation. The fact thrilled me—the magical compression of time and space, the bridging of what felt like myth and my life of autumn raking and primitive video games in Washington, D.C.

The implications were almost unbelievable. If I had just inhaled Caesar's last breath (*Et tu, Brute?*), then I also must have inhaled Beethoven's (*I will hear in heaven*), and Darwin's (*I am not the least afraid to die*). And that of Franklin Delano Roosevelt, and Rosa Parks, and Elvis, and the Pilgrims and Native Americans who attended the first Thanksgiving, and the author of the first suicide note, and even the grandfather I had never met. Ever the descendant of survivors, I imagined Hitler's final breath rising through ten feet of the Führerbunker's concrete roof, thirty feet of German earth, and the trampled roses of the Reich Chancellery, then breaching the Western Front and crossing the Atlantic Ocean and forty years on its way to the second-floor window of my childhood bedroom, where it would inflate me like a deathday balloon.

And if I had swallowed their *last* gasps, I must also have swallowed their *first*, and every breath between. And every breath of everyone. And not only of humans, but all other animals, too: the class gerbil that had died in my family's care, the still-warm chickens my grandmother had plucked in Poland, the final breath of the final passenger pigeon. With each inhale, I absorbed the story of life and death on Earth. The thought granted me an aerial view of history: a vast web woven from one strand. When Neil Armstrong touched boot to lunar surface and said "One small step for man . . . ," he sent out, through the polycarbonate of his visor, into a world without sound, molecules of Archimedes hollering "Eureka!" as he ran naked through the streets of ancient Syracuse, having just discovered that the bathwater displaced by his body was equal to the weight of his body. (Armstrong would leave that boot on the moon, to compensate for the weight of the moon rocks he would bring back.) When Alex, the African grey parrot who was trained to converse at the level of a five-year-old human, uttered his final words—"You be good, see you tomorrow. I love you."—he also exhaled the panting of sled dogs who pulled Roald Amundsen across ice sheets that have since melted and released the cries of exotic beasts brought to the Colosseum to be slaughtered by gladiators. That I had a place in all of that—that I could not escape my place in all of that—was what I found most astonishing.

Caesar's ending was also a beginning: his was among the first recorded autopsies, which is how we know that he was stabbed twenty-three times. The iron daggers are gone. His blood-soaked toga is gone. The Curia of Pompey, in which he was killed, is gone, and the metropolis in which it stood exists only as ruins. The Roman Empire, which once covered two million square miles and encompassed more than 20 percent of the world's population, and whose disappearance was as unimaginable as that of the planet itself, is gone.

It's hard to think of a more ephemeral artifact of a civilization than a breath. But it's impossible to think of a more enduring one.

Despite my recalling so much about it, there was no *Book of Endings*. When I tried to confirm its existence, I found instead *Panati's Extraordinary Endings of Practically Everything and Everybody*, published when I was twelve. It contains Houdini, the fossil record, and many other things that I remembered, but not Caesar's final breath, and not the "Dispute with the Soul," which I must have learned about elsewhere. Those small corrections troubled me—not because they were themselves important, but because my recollections were so clear.

I was further unsettled when I researched the first suicide note and reflected on its title—on the fact that it was titled at all. That we misremember is disturbing enough, but the prospect of being misremembered by those who come after us is deeply upsetting. It remains unknown whether the author of the first suicide note even killed himself. "I opened my mouth to my soul," he writes in the beginning. But the soul has the last word, urging the man to "cling to life." We don't know how the man responded. It is entirely possible that the dispute with the soul resolved with the choice of life, postponing the author's last breath. Perhaps a confrontation with death revealed the most compelling case for survival. A suicide note resembles nothing more closely than its opposite.

No Sacrifice

During World War II, Americans in cities along the East Coast turned off their lights at dusk. They weren't, themselves, in imminent danger; the purpose of the blackout was to prevent German U-boats from using urban backlighting to spot and destroy ships exiting harbor.

As the war progressed, blackouts were practiced in cities across the country, even those far from the coast, to immerse civilians in a conflict whose horrors were out of sight but whose victory would require collective action. On the home front, Americans needed a reminder that life as they knew it could be destroyed, and darkness was one way to illuminate the threat. Civil Air Patrol pilots were encouraged to comb the skies above the Midwest for enemy aircraft, despite the fact that no German fighter plane of the era was capable of flying that far. Solidarity was an important asset, even if such gestures would have been foolish—would have been suicidal—if they were the only efforts made.

World War II would not have been won without homefront actions that had both psychological and tangible impacts:

ordinary people joining together to support the greater cause. During the war, industrial productivity rose by 96 percent. Liberty ships that took eight months to construct at the start of the war were completed in weeks. The SS *Robert E. Peary*—a Liberty ship composed of 250,000 parts weighing fourteen million pounds—was assembled in four and a half days. By 1942, companies that had once manufactured cars, refrigerators, metal office furniture, and washing machines now produced military products. Lingerie factories began making camouflage netting, adding machines were reborn as pistols, and the lung-like bags of vacuum cleaners were transplanted into the bodies of gas masks. Retirees, women, and students entered the workforce—many states changed their labor laws to allow teenagers to work. Everyday commodities like rubber, tin cans, aluminum foil, and lumber were collected for reuse in the war effort. Hollywood studios contributed by producing newsreels, anti-fascist features, and patriotic animated films. Celebrities encouraged the purchase of war bonds, and a few, like Julia Child, became spies.

Congress enlarged the tax base by lowering the minimum taxable income and reducing personal exemptions and deductions. In 1940, 10 percent of American workers paid federal income tax. By 1944, the number approached 100 percent. Top marginal tax rates were raised to 94 percent, while the income that qualified for that rate was reduced by twenty-five-fold.

The government enacted—and Americans accepted—price controls on nylon, bicycles, shoes, firewood, silk, and coal. Gasoline was severely regulated, and a speed limit of thirty-five miles per hour was imposed nationally to reduce gas and rubber consumption. U.S. government posters advocating carpooling declared, "When you ride ALONE you ride with Hitler!"

Farmers—in greatly reduced numbers, and with less

equipment—multiplied their output, and nonfarmers planted "victory gardens," micro-farms in backyards and empty lots. Food was rationed, especially staples like sugar, coffee, and butter. In 1942, the government launched a "Share the Meat" campaign, urging each American adult to limit their weekly meat intake to two and a half pounds. In the U.K., people were eating about half that. (This collective act of belt-tightening led to a general uptick in health.) In July 1942, Disney produced an animated short for the U.S. Department of Agriculture, *Food Will Win the War*, which touted farming as a matter of national security. America had twice as many farmers as the Axis had soldiers. "Their weapons are the panzer forces of food's battle line, farm machinery: battalions of combines; regiments of trucks; divisions of corn pickers, potato diggers, planting machines; columns of milking machines."

On the evening of April 28, 1942, five months after the bombing of Pearl Harbor and well into the war effort in Europe, millions of Americans gathered around their radios to listen to President Roosevelt's fireside chat, in which he gave an update on the state of the war and spoke about the challenges ahead, including what would be asked of citizens:

> Not all of us can have the privilege of fighting our enemies in distant parts of the world. Not all of us can have the privilege of working in a munitions factory or a shipyard, or on the farms or in oil fields or mines, producing the weapons or the raw materials that are needed by our armed forces. But there is one front and one battle where everyone in the United States—every man, woman, and child—is in action, and will be privileged to remain in action throughout this war. That front is

> right here at home, in our daily lives, and in our daily tasks. Here at home everyone will have the privilege of making whatever self-denial is necessary, not only to supply our fighting men, but to keep the economic structure of our country fortified and secure during the war and after the war. This will require, of course, the abandonment not only of luxuries, but of many other creature comforts. Every loyal American is aware of his individual responsibility . . . As I told the Congress yesterday, "sacrifice" is not exactly the proper word with which to describe this program of self-denial. When, at the end of this great struggle, we shall have saved our free way of life, we shall have made no "sacrifice."

It is an extreme burden to be required to give the government 94 percent of your income. It is a significant challenge to have one's food staples rationed. It is a frustrating inconvenience to be able to drive no faster than thirty-five miles per hour. It is slightly annoying to turn off your lights at night.

Despite many Americans' perception of the war as *over there*, a little darkness seems reasonable to ask of citizens who were, after all, largely safe and secure *over here*. How would we regard someone who, in the middle of a great struggle to save not only millions of lives but "our free way of life," deemed turning off his lights too much of a sacrifice?

Of course, the war couldn't have been won *only* with that collective act—victory required sixteen million Americans to serve in the military, more than four trillion dollars, and the armed forces of more than a dozen other countries. But imagine if the war couldn't have been won *without* it. Imagine if preventing Nazi flags from flying in London, Moscow, and Washington,

D.C., required the nightly flipping of switches. Imagine if the remaining 10.5 million Jews of the world could not have been saved without those hours of darkness. How, then, would we regard the self-denial of citizens?

We shall have made no "sacrifice."

Not a Good Story

On March 2, 1955, an African American woman boarded a bus in Montgomery, Alabama, and refused to give up her seat to a white passenger. The average American kid could reenact the scene with feeling, just as surely as she could re-create the first Thanksgiving feast (and know what it meant), throw tea bags from a cardboard boat (and know what it meant), and put on a construction-paper top hat and recite the Gettysburg Address (and know what it meant).

You probably think you know the name of that first woman who refused to move to the back of the bus, but you probably don't. (I didn't until recently.) And that isn't a coincidence or an accident. To some extent, the triumph of the civil rights movement required forgetting Claudette Colvin.

•

The chief threat to human life—the overlapping emergencies of ever-stronger superstorms and rising seas, more severe droughts and declining water supplies, increasingly large ocean dead zones, massive noxious-insect outbreaks, and the daily disappearance of forests and species—is, for most people, not a good story. When the planetary crisis matters to us at all, it has the quality of a war being fought *over there*. We are aware of the existential stakes and the urgency, but even when we know that a war for our survival is raging, we don't feel immersed in it. That distance between awareness and feeling can make it very difficult for even thoughtful and politically engaged people—people who *want* to act—to act.

When the bombers are overhead, as they were in wartime London, it goes without saying that you will turn off all your lights. When the bombing is off the coast, it doesn't go without saying, even if the ultimate danger is just as great. And when the bombing is across an ocean, it can be hard to believe in the bombing at all, even though you know it is happening. If we don't act until we feel the crisis that we rather curiously call "environmental"—as if the destruction of our planet were merely a context—everyone will be committed to solving a problem that can no longer be solved.

Compounding the *over there* quality of the planetary crisis is a fatigue of the imagination. It is exhausting to contemplate the complexity and scale of the threats we face. We know climate change has something to do with pollution, something to do with carbon, ocean temperatures, rainforests, ice caps . . . but most of us would find it difficult to explain how our individual and collective behavior is boosting hurricane winds by almost thirty miles per hour or contributing to a polar vortex that makes Chicago colder than Antarctica. And we find it hard to remember how much the world has already changed: we don't balk at proposals like the construction of a ten-mile-long seawall around Manhattan, we accept increased insurance premiums, and extreme weather—forest fires encroaching on metropolises, annual "thousand-year floods," record deaths from record heat waves—is now just weather.

In addition to it not being an easy story to tell, the planetary crisis hasn't proved to be a *good* story. It not only fails to convert us, it fails to interest us. To captivate and to transform are the most fundamental ambitions of activism and art, which is why climate change, as subject matter, fares so poorly in both realms. Revealingly, the fate of our planet occupies an even smaller place in literature than it does in the broader cultural conversation, despite most writers considering themselves especially sensitive to the underrepresented truths of the world. Perhaps that's because writers are also especially sensitive to what kinds of stories "work." The narratives that persist in our culture—folktales, religious texts, myths, certain passages of history—have unified plots, sensational action between clear villains and heroes, and moral conclusions. Hence the instinct to present climate change—when it is presented at all—as a dramatic, apocalyptic event in the future (rather than a variable, incremental process occurring over time), and to paint the fossil fuel industry as the embodiment of destruction (rather than one of several forces that require our attention). The planetary crisis—abstract and eclectic as it is, slow as it is, and lacking in iconic figures and moments—seems impossible to describe in a way that is both truthful and enthralling.

•

Claudette Colvin was the first woman to be arrested for refusing to change bus seats in Montgomery. Rosa Parks, whose name most of us know, didn't come along for another nine months. And when it was Parks's moment to resist bus segregation, she was not, as the story goes, simply an exhausted seamstress returning home at the end of a long day. She was a civil rights activist (the secretary of her local NAACP chapter) who had attended social justice workshops, lunched with influential lawyers, and participated in strategizing the movement's tactics. Parks was forty-two, married, and from a respected family. Colvin was fifteen, pregnant with the child of an older, married man, and from a poor family. Civil rights leaders—including Parks herself—regarded Colvin's biography as too imperfect, and her character too volatile, for her to be the hero of the emerging movement. It wouldn't make a sufficiently good story.

Would Christianity have spread if instead of being crucified on a cross, Jesus had been drowned in a bath? Would Anne Frank's diary be so widely read if she had been a middle-aged man hidden behind a cupboard, rather than a hauntingly beautiful girl behind a bookcase? To what extent was the course of history influenced by Lincoln's stovepipe hat, Gandhi's loincloth, Hitler's mustache, Van Gogh's ear, Martin Luther King's cadence, the fact that the Twin Towers were the two most easily drawn buildings on the planet?

The story of Rosa Parks is both a true episode from history and a fable created to make history. Like the iconic photographs of the soldiers raising the flag at Iwo Jima, the couple kissing in Robert Doisneau's *Le baiser de l'hôtel de ville*, and the milkman moving through the rubble of bombed-out London, the photo of Rosa Parks on the bus was staged. It is a sympathetic journalist, not an aggravated segregationist, seated behind her. And as she later acknowledged, what happened wasn't quite as simple—as memorable—as a tired woman being told to move from the front of the bus to the back. But she embodied the most inspiring version of events because she understood the power of narrative. Parks was brave for being the hero of her story, but heroic for being one of its authors.

History not only makes a good story in retrospect; good stories *become* history. With regard to the fate of our planet—which is also the fate of our species—that is a profound problem. As the marine biologist and filmmaker Randy Olson put it, "Climate is quite possibly *the* most boring subject the science world has ever had to present to the public." Most attempts to narrativize the crisis are either science fiction or dismissed as science fiction. There are very few versions of the climate change story that kindergartners could re-create, and there is no version that would move their parents to tears. It seems fundamentally impossible to pull the catastrophe from over there in our contemplations to right here in our hearts. As Amitav Ghosh wrote in *The Great Derangement*, "The climate crisis is also a crisis of culture, and thus of the imagination." I would call it a crisis of belief.

Know Better, No Better

In 1942, a twenty-eight-year-old Catholic in the Polish underground, Jan Karski, embarked on a mission to travel from Nazi-occupied Poland to London, and ultimately America, to inform world leaders of what the Germans were perpetrating. In anticipation of his journey, he met with several resistance groups, accumulating information and testimonies to bring to the West. In his memoir, he recounts a meeting with the head of the Jewish Socialist Alliance:

> The Bund leader came up to me in silence. He gripped my arm with such violence that it ached. I looked into his wild, staring eyes with awe, moved by the deep, unbearable pain in them.
>
> "Tell the Jewish leaders that this is no case for politics or tactics. Tell them that the Earth must be shaken to its foundation, the world must be aroused. Perhaps then it will wake up, understand, perceive. Tell them that they must find the strength and courage to make sacrifices no other statesmen have ever had to make,

> sacrifices as painful as the fate of my dying people, and as unique. This is what they do not understand. German aims and methods are without precedent in history. The democracies must react in a way that is also without precedent, choose unheard-of methods as an answer . . .
>
> "You ask me what plan of action I suggest to the Jewish leaders. Tell them to go to all the important English and American offices and agencies. Tell them not to leave until they obtain guarantees that a way has been decided upon to save the Jews. Let them accept no food or drink, let them die a slow death while the world is looking on. Let them die. This may shake the conscience of the world."

After surviving as perilous a journey as could be imagined, Karski arrived in Washington, D.C., in June 1943. There, he met with Supreme Court Justice Felix Frankfurter, one of the great legal minds in American history, and himself a Jew. After hearing Karski's accounts of the clearing of the Warsaw Ghetto and of exterminations in the concentration camps, after asking him a series of increasingly specific questions ("What is the height of the wall that separates the ghetto from the rest of the city?"), Frankfurter paced the room in silence, then took his seat and said, "Mr. Karski, a man like me talking to a man like you must be totally frank. So I must say I am unable to believe what you told me." When Karski's colleague pleaded with Frankfurter to accept Karski's account, Frankfurter responded, "I didn't say that this young man is lying. I said I am unable to believe him. My mind, my heart, they are made in such a way that I cannot accept it."

Frankfurter didn't question the truthfulness of Karski's story. He didn't dispute that the Germans were systematically murdering the Jews of Europe—his own relatives. And he didn't respond that while he was persuaded and horrified, there was nothing he could do. Rather, he admitted not only his inability to believe the truth but his awareness of that inability. Frankfurter's conscience was not shaken.

Our minds and hearts are well built to perform certain tasks, and poorly designed for others. We are good at things like calculating the path of a hurricane, and bad at things like deciding to get out of its way. Because we evolved over hundreds of millions of years, in settings that bear little resemblance to the modern world, we are often led to desires, fears, and indifferences that neither correspond nor respond to modern realities. We are disproportionately drawn to immediate and local needs—we crave fats and sugars (which are bad for people who live in a world of their ready availability); we hyper-vigilantly watch our children on jungle gyms (despite the many greater risks to their health that we ignore, like overfeeding them fats and sugars)—while remaining indifferent to what is lethal but *over there*.

In a recent study, the UCLA psychologist Hal Hershfield found that when subjects were asked to describe their future selves—even a mere ten years from now—their brain activity on fMRI scans bore more resemblance to what appeared when they described strangers than to what appeared when they described their current selves. When subjects were shown images of themselves in which their appearance had been digitally aged, however, this disparity changed, and so did their behavior. Asked to allocate a thousand dollars among four options—a gift for a loved one, a fun event, a checking account, or a retirement fund—

subjects who saw their aged avatars put nearly twice as much money into their retirement accounts as subjects who didn't.

It has been widely demonstrated that emotional responses are heightened by vividness. Researchers have described a number of "sympathy biases" that generate concern: the identifiable-victim effect (the ability to visualize the details of the suffering), the in-group effect (the suggestion of social proximity to the suffering), and the reference-dependent sympathy effect (the presentation of the victim's condition as not merely dreadful but worsening). One group of researchers conducted a direct-mail fundraising experiment with about two hundred thousand potential donors. If the mailing featured a named individual as opposed to an unnamed group, donations increased by 110 percent. If the donor and the target belonged to the same religion, donations increased by 55 percent. If the target's poverty was presented as newfound instead of chronic, donations increased by 33 percent. Combining all these tactics led to a 300 percent increase in donations.

The problem with the planetary crisis is that it runs up against a number of built-in "apathy biases." Although many of climate change's accompanying calamities—extreme weather events, floods and wildfires, displacement and resource scarcity chief among them—are vivid, personal, and suggestive of a worsening situation, they don't feel that way in aggregate. They feel abstract, distant, and isolated rather than like beams of an ever-strengthening narrative. As the journalist Oliver Burkeman put it in *The Guardian*, "If a cabal of evil psychologists had gathered in a secret undersea base to concoct a crisis humanity would be hopelessly ill-equipped to address, they couldn't have done better than climate change."

So-called climate change deniers reject the conclusion that

97 percent of climate scientists have reached: the planet is warming because of human activities. But what about those of us who say we accept the reality of human-caused climate change? We may not think the scientists are lying, but are we able to believe what they tell us? Such a belief would surely awaken us to the urgent ethical imperative attached to it, shake our collective conscience, and render us willing to make small sacrifices in the present to avoid cataclysmic ones in the future.

Intellectually accepting the truth isn't virtuous in and of itself. And it won't save us. As a child, I was often told "you know better" when I did something I shouldn't have done. *Knowing* was the difference between a mistake and an offense.

If we accept a factual reality (that we are destroying the planet), but are unable to *believe* it, we are no better than those who deny the existence of human-caused climate change—just as Felix Frankfurter was no better than those who denied the existence of the Holocaust. And when the future distinguishes between these two kinds of denial, which will appear to be a grave error and which an unforgivable crime?

Be Leaving, Believing, Be Living

A year before Karski journeyed from Poland to inform the world that the Jews of Europe were being slaughtered, my grandmother fled her Polish village to save her life. She left behind four grandparents, her mother, two siblings, cousins, and friends. She was twenty years old and knew only what everyone else knew: the Nazis were pushing east into Soviet-occupied Poland and were only days away. Asked why she left, she would say, "I felt I had to do something."

My great-grandmother, who would be shot at the edge of a mass grave while holding her stepdaughter, watched my grandmother pack her things. They didn't speak. That silence was their final exchange. Knowing no less than her daughter, she didn't feel that she had to do something. Her knowledge was only knowledge.

My grandmother's younger sister, who would be shot trying to trade a trinket for something to eat, followed my grandmother out of the house that day. She took off her only pair of shoes and gave them to my grandmother. "You're so lucky to be leaving," she said. I've been told that story many times. As a child I heard it as "You're so lucky believing."

Maybe it is just luck. If a few factors had been different around the time that my grandmother left—if she had been ill, or if she had just fallen in love with someone—maybe she would not have been lucky to be leaving. Those who stayed weren't any less brave, intelligent, resourceful, or afraid of dying. They just didn't believe that what was coming would be so different from what had already come many times. Belief can't be willed into being. And you can't force someone to believe, not even with better and louder and more virtuous arguments, not even with irrefutable evidence. As the filmmaker Claude Lanzmann puts it in his spoken prologue to *The Karski Report*, a documentary about Karski's visit to America:

> What is knowledge? What can information about a horror, a literally unheard-of one, mean to the human brain, which is unprepared to receive it because it concerns a crime that is without precedent in the history of humanity? . . . Raymond Aron, who had fled to London, was asked whether he knew what was happening at that time in the East. He answered: I knew, but I didn't believe it, and because I didn't believe it, I didn't know.

I sometimes daydream about going from house to house in my grandmother's shtetl, grabbing the faces of those who would stay, and screaming, "You have to do something!" I have this daydream in a house that I *know* consumes multiples of my fair share of energy and I *know* is representative of the kind of voracious lifestyle that I *know* is destroying our planet. I am capable of imagining one of my descendants daydreaming about grabbing my face and screaming, "You have to do something!" But I am incapable of the belief that would move me to do something. So I know nothing.

The other morning, on the drive to school, my ten-year-old son looked up from the book he was reading and said, "We are so lucky to be living."

One piece of knowledge I don't have: how to square my own gratitude for life with behavior that suggests an indifference to it.

My grandmother took her winter coat when she left home, even though it was June.

Hysterical

One summer night in 2006, eighteen-year-old Kyle Holtrust was riding his bicycle against traffic on the east side of Tucson when a Chevy Camaro struck him and dragged him beneath it for thirty feet. A witness in a nearby truck, Thomas Boyle, Jr., leaped from the passenger seat and ran over to help. Flooded with adrenaline, he gripped the frame of the Camaro and lifted its front end, holding it aloft for forty-five seconds while Holtrust was pulled free. When explaining why he did what he did, Boyle said, "I would be such a horrible human being to watch someone suffer like that and not even try to help . . . All I could think is, what if that was my son?" He felt he had to do something.

But when asked *how* he did what he did, he was at a loss: "There's no way I could lift that car right now." The world record for a dead lift is 1,102 pounds. A Camaro weighs between 3,300 and 4,000 pounds. Boyle, who was not a weight lifter, exhibited what is called "hysterical strength"—a physical feat, performed in a life-or-death situation, that exceeds what is usually considered possible.

One amazing person lifted the car off Holtrust's body, but then many people pulled their cars to the side of the road to make the ambulance's journey quicker. They were every bit as important in saving the young man's life, but we don't think of their acts as exceptional. To lift a car into the air is the most one can do. To move your car to the side when an ambulance appears is the least one can do. Kyle's life depended on both.

When I was in grade school, police officers and firefighters gave annual presentations intended to inspire civic awareness and responsibility and to educate us about what to do in dangerous situations. I remember a fireman telling us that every time we saw an ambulance, we should imagine it carrying someone we love. What a miserable thought to deposit in a child's head! Especially because it doesn't make the right connection. We don't get out of the way of an approaching ambulance because a loved one might be in it. And we don't get out of the way because it's the law. We do it because it is *what we do*. Making way for an ambulance is one of those social norms—like waiting in lines and putting garbage in a garbage can—that is so ingrained in our culture we don't even notice it.

Norms can change, and they can be ignored. In Moscow in the early 2010s, there was a rash of "ambulance taxis"—vans made to look like emergency vehicles on the outside but outfitted with luxurious interiors and rented out for upwards of two hundred dollars per hour for the purpose of beating the city's infamously bad traffic. It's hard to imagine anyone who isn't inside one of those vehicles being okay with them. They are an affront—not because we are being taken advantage of as individuals (most of us will never be passed by such a vehicle) but because they violate our willingness to sacrifice for the collective good. They exploit our best impulses. Home-front blackouts led to looting during World War II, and food rationing to forgery and theft. In London, when a Piccadilly nightclub suffered a direct hit by the Luftwaffe, rescuers had to fend off those trying to take jewelry from the dead.

But those are extreme examples. Almost always, our conventions and the identities they form are subtle to the point of being invisible. Sure, we don't drive around in fake ambulances, but many of the ways we now live will look as bad (and far worse) to our descendants. The word "ambulance" is written in reverse on the hoods of ambulances so that it can be read in the rearview mirrors of drivers in front of them. You could say that the word is written for the future—for cars that are ahead on the road. Just as someone in an ambulance can't see the word "ambulance," we can't read the history we're creating: it's written in reverse, to be read in a rearview mirror by those who aren't yet born.

The word "emergency" derives from the Latin *emergere*, which means "to arise, bring to light."

The word "apocalypse" derives from the Greek *apokalyptein*, which means "to uncover, to reveal."

The word "crisis" derives from the Greek *krisis*, meaning "decision."

Encoded into our language is the understanding that disasters tend to expose that which was previously hidden. As the planetary crisis unfolds as a series of emergencies, our decisions will reveal who we are.

Different challenges require, and inspire, different reactions. Alarm is an appropriate response to a person pinned beneath a car, but someone who abandons his otherwise beautiful home because of a tiny leak is alarmingly overreactive. What does the condition of the planet require, and what does it inspire? And what happens if it doesn't inspire what it requires—if we reveal ourselves to be people who put flashing lights atop our vehicles to avoid traffic but won't turn off the lights in our homes to avoid destruction?

Away Games

Despite the numerous observed instances of hysterical strength, it has never been demonstrated in a controlled setting, because it would be unethical to create the necessary conditions. But even beyond the witnessed cases, there are reasons to assume it is a real phenomenon, including the effects of electrical charges on muscles (which demonstrate a strength that far exceeds what can be willed) and the performance of athletes at the most important competitions. It is no coincidence that the great majority of world records are set at the Olympics, when the viewing audience is vastly larger than that of any other competition and the stakes are much higher. Athletes are able to try harder because they care more.

Across sports, individuals and teams win more often when competing at home. (Not only are the majority of world records set at the Olympics, but the host country almost always overperforms.) Some portion of this can be explained by getting a good night's rest in one's own bed the night before, eating home-cooked food, and playing on a familiar field. Some can be explained by referee bias toward the home team. But the greatest advantage may be delivered by the supporters: playing in a stadium of one's fans generates confidence and offers a powerful incentive to win. A study of Germany's Bundesliga soccer league demonstrated that home-field advantage is greater in stadiums where the soccer field is not surrounded by a track than in those where it is. The closer the fans are to the field, the more their presence is felt—the more home *feels* like home.

It is natural to assume that if we are to summon the necessary will to meet the planetary crisis, we will have to summon the necessary care. We will need to regard Earth as our only home—not idiomatically, and not intellectually, but viscerally. As the Nobel Prize–winning psychologist Daniel Kahneman, who pioneered the understanding that our minds have a slow (deliberative) and a fast (intuitive) mode, put it, "To mobilize people, this has to become an emotional issue." If we continue to experience the struggle to save our planet as a midseason away game, we will be doomed.

Clearly, facts aren't enough to mobilize us. But what if we can't summon and sustain the necessary emotions? I've wrestled with my own responses to the planetary crisis. It feels obvious to me that I care about the fate of the planet, but if time and energy invested are expressions of caring, it's undeniable that I care more about the fate of a specific baseball team on the planet, my childhood-hometown Washington Nationals. It feels obvious to me that I am not a climate change denier, but it is undeniable that I behave like one. I would let my kids skip school to participate in the wave at opening day of baseball season, but I do virtually nothing to resist a future in which our home city is underwater.

When researching this book, I was often shocked by what I learned. But I was rarely moved by it. When I was moved, the feeling was transient, and it was never deep enough or durable enough to change my behavior over time. Even the reporting that terrified me, like David Wallace-Wells's chilling essay "The Uninhabitable Earth"—the most-read article in the history of *New York* magazine at its time of publication—failed to shake my conscience or to take permanent residence there. This is not the fault of the essay, which is not only revelatory but clever, and pleasurable—in the way that only a nonfiction apocalyptic prophecy can be. It is the fault of the subject. It is excruciatingly, tragically difficult to talk about the planetary crisis in a way that is believed.

Thomas Boyle, Jr., didn't need information to inspire him to lift the Camaro off Kyle Holtrust; he needed feeling: "All I could think is, what if that was my son?" But what if the emotional connection weren't as strong? Would he have lifted the car—would he have been able to, would he have tried to—if it had been more difficult to imagine Holtrust as his child? If Holtrust had been a different race or age? What if Boyle had been watching a simulcast of the accident on a screen and been told that deadlifting three thousand pounds would save a victim halfway around the world? Despite the loving relationships most people have with their pets, and the frequency with which animals get struck by cars, there has never been an observed instance of an individual lifting a car off a trapped dog or cat. Our bodies have limits, and so do our emotions. But what if our emotional limits cannot be exceeded?

Writing the Word "Fist"

The last time I checked on my roof was too long ago to say how long ago. It is out of mind because it is out of sight—I literally can't see its condition, and unlike a water stain on the ceiling, which is aesthetically unpleasant, a decrepit roof is no eyesore or embarrassment. Even if I managed to examine it, as a layperson I probably wouldn't know it was in need of maintenance until it was in need of replacement. The prospect of having to replace my roof has discouraged me from determining whether I need to.

My younger son had a nightmare while I was taking a shower the other night. I heard his cry through the water, the glass door, and the three walls that separated us. By the time I got to his bed, he had already returned to a peaceful sleep. His lavishly decorated bedroom is beneath a roof that might be deteriorating.

Hysterical strength could explain my ability to hear his quiet cries, but what deficiency allows me to ignore the precarious roof, and the precarious sky above it? I would bet that at some point, every one of the Jews in my grandmother's village swatted a fly that had landed on their skin. Whatever it is that allows me to ignore my roof and the climate is the same thing that allowed so many of them to stay behind when they knew the Nazis were coming. Our alarm systems are not built for conceptual threats.

I was in Detroit when Hurricane Sandy was about to hit the Eastern Seaboard. All the flights back to New York had been canceled, and it wouldn't be possible to get on a plane for the next several days. The prospect of not being with my family was intolerable to me. There was nothing to be done at home—we had plenty of bottled water and nonperishable food in the pantry, flashlights with fresh batteries—but I had to be there. I found the last rental car in the area and hit the road at eleven that night. Twelve hours later, I was driving through the front edge of the storm. The wind and rain made progress almost impossible. The final hour took four hours. The kids were sleeping when I got home. I called my parents, as I had promised I would, and my mother told me, "You're a great father."

I had driven sixteen hours to get home simply to be there. In the days, months, and years after, I did virtually nothing to lessen the chances of another superstorm pummeling my city. I barely even entertained the question of what I could do.

It felt good to make that drive. Being there, doing nothing, felt good. It felt good to hear my mother's praise for my parenting and, when they came downstairs, to see my children's relief at my presence. But what kind of father prioritizes feeling good over doing good?

I was a boy when I learned why the word "ambulance" is written in reverse. I loved the explanation. But now I'm older, and there's something I can't figure out: Is there anyone alive who would see an ambulance in their rearview mirror—the bright lights spinning, the sirens blaring—and require the word "ambulance" to identify it? Isn't it like a boxer writing the word "fist" on his boxing glove?

I run to soothe a nightmare in my son's head but do almost nothing to prevent a nightmare in the world. If only I could perceive the planetary crisis as a call from my sleeping child. If only I could perceive it as exactly what it is.

Sometimes a fist needs the word "fist" written across it. Hurricane Sandy battered our home and our city. We received those punches without being able to identify them as punches; to most of us, they were just weather. Journalists, news anchors, politicians, and scientists were wary of identifying it as a product of climate change until there was evidence of the kind of irrefutability that will never come. And anyway, what does one do with weather but accept it?

I want to care about the planetary crisis. I think of myself, and want to be thought of, as someone who cares. Just as I think of myself, and want to be thought of, as a great father. Just as I think of myself, and want to be thought of, as someone who cares about civil liberties, economic justice, discrimination, and animal welfare. But these identities—which I flaunt with exhibitionist conscientiousness and dinner-party Op-Eding—inspire responsibility less often than they get me off the hook. They don't reflect truths so much as offer ways to evade them. They are not identities at all—only identifiers.

The truth is I don't care about the planetary crisis—not at the level of belief. I make efforts to overcome my emotional limits: I read the reports, watch the documentaries, attend the marches. But my limits don't budge. If it sounds like I'm protesting too much or being too critical—how could someone claim indifference to the subject of his own book?—it's because you also have overestimated your commitment while underestimating what is required.

In 2018, despite knowing more than we've ever known about

human-caused climate change, humans produced more greenhouse gases than we've ever produced, at a rate three times that of population growth. There are tidy explanations—the growing use of coal in China and India, a strong global economy, unusually severe seasons that required spikes in energy for heating and cooling. But the truth is as crude as it is obvious: we don't care.

So now what?

Sticks

Just as our descendants won't distinguish between those who denied the science of climate change and those who behaved as if they did, neither will they distinguish between those who felt a deep investment in saving the planet and those who simply saved it. It might be that we cannot muster strong feelings about our home. It might also be that we don't need to. In this case, feelings could impede progress rather than facilitate it.

The first photographic portrait of a human was taken in 1839—it was a selfie. Robert Cornelius set up a box fitted with a lens from an opera glass in back of his family's lamp and chandelier store in Philadelphia. He removed a light-blocking cap from the lens, ran into the frame, remained still for more than a minute, and then ran back and replaced the cap. A little less than two centuries later, ninety-three million selfies are taken every day by Android users alone. Researchers recently identified a condition defined by an urge to take selfies and upload them to social media at least six times a day. They named it "chronic selfitis."

If a cabal of evil psychologists had concocted climate change as the perfect catastrophe to destroy our species, they might have thrown in MSNBC, social media, and hybrid cars for good measure, each of which can provide a feeling of engagement at the expense of engagement, in much the same way that selfies can make us feel present at the expense of being present.

Explaining the rise of MSNBC, the Republican strategist Stuart Stevens said, "I think there are a lot of people out there who are dramatically troubled by the direction of the country, and they would like to be reminded that (*a*) they're not alone, and, (*b*) there is an alternative." But loneliness isn't the problem; the direction of the country is. And being alone together is not an alternative direction, just as a cancer-support group does not shrink a tumor. It's probably true that viewers of MSNBC are sometimes inspired to give money to progressive candidates, and perhaps there is someone out there whose politics were changed, rather than their loneliness assuaged, by Rachel Maddow. It's certainly true that a hybrid car gets better mileage than a traditional gas car. But primarily, these things make us feel better. And it can be dangerous to feel better when things are not getting better.

A recent study published in *Environmental Science and Technology* examined 108 scenarios for adoption of hybrid and fully electric vehicles over the next three decades, considering such variables as oil and gas prices, battery costs, government incentives for alternative fuels, and possible emissions caps. It found that because decreases in tailpipe emissions are largely offset by the increased electricity generation needed to charge car batteries, "the model results do not demonstrate a clear and consistent trend toward lower system-wide emissions." While that conclusion might be debatable, what isn't is that an average person's vehicle emissions are no more than 20 percent of their total carbon emissions. Even managing to live car-free—a far more significant action than switching to a Prius—would only be a start. We need to use cars far less, but we need to do far more than just that. Too often, the feeling of making a difference doesn't correspond to the difference made—worse, an inflated sense of accomplishment can relieve the burden of doing what actually needs to be done.

Do the children getting vaccines paid for by Bill Gates really care if he feels annoyed when he gives 46 percent of his vast wealth to charity? Do the children dying of preventable diseases really care if Jeff Bezos feels altruistic when he donates only 1.2 percent of his even vaster wealth?

If you found yourself in the back of an ambulance, would you rather have a driver who loathes his job but performs it expertly or one who is passionate about his job but takes twice as long to get you to the hospital?

To save the planet, we need the opposite of a selfie.

A Wave

Honeybees perform a wave to ward off predatory hornets. One after the other, individual bees momentarily flip their abdomens upward, creating an undulating pattern across the nest—the phenomenon is called "shimmering." The collective fends off the threat, something no individual bee could do on its own.

For every story of an individual lifting a car off someone trapped beneath it, there are a hundred stories of groups of people lifting a car off someone who's trapped. (And while there are no stories of a person lifting a car off a pet, there are many stories of groups of people doing so.) Someone trapped beneath a car does not distinguish between the astonishing act of the individual and the smaller collective efforts of individuals acting together.

Einstein is quoted as saying, "If the bee disappeared off the face of the earth, man would have four years left to live." He almost certainly didn't say that, and the statement almost certainly isn't true. Just as the widely cited statistic that one-third of all food crops rely on bee pollination isn't accurate. But it *is* the case that bee populations have been collapsing globally because of changing temperatures (among other things, like pesticides, monoculture, and habitat loss from industrial agriculture), and not only will the effects be profound but they are already being felt—determining what kinds of crops can be grown, how they are priced, and how they are farmed.

From China to Australia to California, fruit and nut farmers will often rent bees trucked in from hundreds of miles away to pollinate their trees. And in areas where human labor is less expensive than bee labor—a thought worth pausing on—the trees are pollinated by hand. Workers swarm the fields. Using long sticks with chicken feathers and cigarette filters on one end, they painstakingly transfer pollen from bottles around their necks into the stigma of every flower. A photographer who documented this process said, "On the one hand it's a story about the human toll on the environment, while on the other it shows our ability to be more efficient in spite of it all."

Really? Can any sense of the word "efficient" describe a situation where humans are required to do the work of bees? Is there anything resembling an inspiring, or even acceptable, *other hand*?

Selfie sticks perfectly symbolize the supremacy of social performance—*look at me doing something*. Pollen sticks perfectly symbolize our planetary crisis—*look what happens when no one does anything*. It might not be the case that the selfie stick inevitably evolves into a pollen stick, but it won't be possible to move away from the latter without moving away from the former.

Hold these two images in your mind: an individual lifting a car off someone trapped beneath it, and hundreds of human workers painstakingly placing pollen into flowers. Are those our only options when responding to a crisis? Hysterical strength or hysterical weakness?

No, there is a third option.

I have never started a wave at a baseball game. Waves do not require any more initiative than participation.

I have never experienced a wave reaching me at the precise moment of my feeling roused by enthusiasm. Waves do not require feeling; they generate feeling.

I have never resisted a wave.

Feel Like Acting, Act Like Feeling

Ninety-six percent of American families gather for a Thanksgiving meal. That is higher than the percentage of Americans who brush their teeth every day, have read a book in the last year, or have ever left the state in which they were born. It is almost certainly the broadest collective action—the largest wave—in which Americans partake.

If Americans had set a goal to eat as many turkeys as possible on one day, it's awfully hard to imagine how we could surpass the forty-six million that are consumed on the third Thursday of November every year. If President Roosevelt had asked us to eat turkeys to support the war effort, if President Kennedy had inspired a moon shot of turkey consumption, I doubt we would have eaten this many. If turkey meals were given out for free on every street corner, I don't believe more than forty-six million would be eaten. Not even if people were *paid* to eat turkey. If there were a law obligating Americans to have Thanksgiving meals, the number of people celebrating Thanksgiving would drop.

In his landmark book *The Gift Relationship: From Human Blood to Social Policy*, the social scientist Richard Titmuss argues that paying blood donors risks having the opposite of its intended effect, because it undermines the most important motivation: altruism. A recent study by the Stockholm School of Economics sought to test Titmuss's theory, and indeed it discovered that for some populations—the findings were particularly dramatic for women—the supply of blood donors can decrease by as much as half when a monetary payment is included.

If you celebrate Thanksgiving—or Christmas, or Passover, or any collective commemoration—do you do so because there are external incentives, like a law or monetary compensation? Because you are spontaneously moved? Or because, like allowing an ambulance to pass or rising when the wave approaches at a baseball game, *it is there* and it is *what you do*? Thanksgiving certainly offers pleasures (a good meal, time with family) and frustrations (the hassle of travel, time with family), but for most people, those factors don't determine the decision to celebrate.

How many people actually *decide* to celebrate Thanksgiving every year? If the possibility of abstaining were built into the culture—as it is for many national secular holidays, like the Fourth of July—would 96 percent really make the same choice? We arrive at the table not because of feelings but because Thanksgiving is on the calendar, and because we've never skipped it before. We do it because we do it. Often, merely participating in an activity produces the feeling that was meant to inspire it in the first place.

There was a study conducted at the Magh Mela, a Hindu festival held in Allahabad, India, and considered to be one of the world's largest collective events. The control subjects—"comparable others"—did not attend and reported no change in their spiritual identities a month after the event. But pilgrims who had participated "exhibited heightened social identification as a Hindu and increased frequency of prayer rituals." An altogether different study found that couples who were asked to cuddle for longer than they normally would after having sex reported greater satisfaction with their relationships than the control group. "Engaging in longer and more satisfying post-sex affection over the course of the study was associated with higher relationship and sexual satisfaction three months later," the researchers concluded.

While it's true that people celebrate Thanksgiving to express gratitude, and participate in religious festivals to express religious identity, and cuddle to express affection, the initial motivation doesn't always need to be strong, or even present at all. Motivation can beget action, but—more remarkably—action can also beget motivation. We don't trek to the desert to look up at the stars because we're feeling spiritual. We feel spiritual because we are in the desert looking at the stars. We don't fight lines at airports and travel thousands of miles for Thanksgiving dinner because we feel especially close to our families in the third week of November. We feel that special closeness because of the journey and meal.

After the grocery store Pay and Save put green arrows on the floor leading to fruit and vegetable aisles, 90 percent of shoppers followed the path, and sales of fresh produce skyrocketed.

In countries where citizens have to opt in to organ donation, about 15 percent do so. In countries where they have to opt out—where the default is organ donation—the number of organ donors rises to about 90 percent.

By nudging men to aim at the right place, playful stickers—houseflies, targets, the New England Patriots' logo—applied near urinal drains reduce spillage by as much as 80 percent.

While it is probably the case that if the celebration of Thanksgiving were legally enforced, fewer people would celebrate it, it is certainly the case that if the celebration of Thanksgiving were not facilitated by being a national holiday, fewer people would celebrate it. The collective action occurs because the structure encourages it—our amorphous, un-urgent emotions about Thanksgiving need a scaffolding.

•

About 37 percent of registered voters voted in the 2014 midterm elections in the United States. In the 2016 presidential election—widely referred to as "the most important election of our time"—about 60 percent voted. Why is there near-unanimous participation in the collective action of Thanksgiving while so few people participate in American democracy? Each requires some work, and each offers a deep gratification. But only one of them shapes the world for the next four years. We have no trouble celebrating history together, but we find it difficult to participate in its creation.

Unlike Thanksgiving, Election Day is not a national holiday. Although the events often occur weeks apart, and although the latter is far more practically consequential than the former, significantly more people show up at Thanksgiving than do at the polls. Thanksgiving is inviting. For many, voting is prohibitive. Most people would describe their experience of Thanksgiving as sitting at a table and enjoying a leisurely meal with loved ones. Most people would describe voting as standing in a long line among strangers, often in inclement weather, worrying about being late to work, or worrying about being late to dinner, and then worrying about whether they filled out the bizarrely complicated ballot properly.

Of course, there is an alternative. We could make Election Day a national holiday, giving everyone the day off from work and school. We could allow people to vote online, as we allow them to pay taxes online. We could greatly simplify the ballot, show images of the candidates beside their names . . .

Various architectures exist to encourage the celebration of certain values and the consumption of certain foods on Thanksgiving. Architectures also exist to discourage voting.

Some events—seeing a teenager trapped under a car, hearing a child crying out in his sleep, feeling an insect land on your skin, competing at an Olympic event, participating in military combat—generate feelings that facilitate actions. Yet many events equally, and often more urgently, require actions that they don't inspire. Conceptual events—the Nazis approaching your village, a national observance of gratitude, an offshore war, a presidential election, climate change—require structures that facilitate actions that generate feelings.

Building a new structure requires architects, and often it requires dismantling the existing structures in the way, even if we've grown so accustomed to seeing them that we no longer see them at all.

Where Do Waves Begin?

"When, at the end of this great struggle, we shall have saved our free way of life, we shall have made no 'sacrifice.'" Americans heard those disembodied words through their radios; Roosevelt delivered them from a wheelchair. History's most public polio patient was also its most private. He never denied having lost the use of his legs, but he carefully stage-managed his political appearance: photographers who took pictures of him in a wheelchair were banned from the White House press corps; he rarely entered or exited a car in public; he wore steel braces to support his legs when standing. If you've ever watched a video of Roosevelt delivering a speech—perhaps the "Infamy" speech to Congress—you've probably noticed that his head gestures almost spastically. His chin is substituting for his hands, which are gripping the podium to keep him upright.

Despite his privacy, Roosevelt was instrumental in the development of a polio vaccine. In 1938, he helped create the organization that would come to be known as the March of Dimes, which became the primary funding source for polio research. One recipient of that funding was Jonas Salk. In 1952, after successfully inoculating thousands of monkeys with his unorthodox "killed virus" vaccine, Salk began human testing—his first patients were himself, his wife, and their three sons. Two years later, he initiated the clinical trial, which would become the largest public health experiment in the country's history. Despite there being no guarantee that the vaccine was safe, nearly two million people became "polio pioneers." On April 12, 1955—a decade, to the day, after Roosevelt's death—the results of the trials were made public. The vaccine was "safe, effective, and potent." Jonas Salk had cured polio.

•

When a social norm changes quickly, it allows people—*releases* them—to act. But as with a wave at a baseball game, even if the participants are eager to join, collective actions need to be set in motion. For more than two hundred years after the first Thanksgiving meal, different colonies, and then states, commemorated different Thanksgivings. They were celebrated on different days (often in different seasons); some involved feasts of regionally specific foods, and some involved fasting. George Washington declared a Thanksgiving Day in February 1795. John Adams declared one in 1798 and one in 1799. Thomas Jefferson chose not to declare any. It wasn't until 1863—the middle of the Civil War—that Abraham Lincoln, in an effort to unify the fractured nation, proclaimed the last Thursday of every November a national holiday. The Thanksgiving that we celebrate today is meant to commemorate a feast shared between the Plymouth colonists and Wampanoag Indians in 1621, but when Lincoln first proposed the holiday in a speech, he emphasized general gratitude that "harmony has prevailed everywhere except in the theatre of military conflict." Whatever his reasons, by codifying the holiday and making it easy to celebrate, Lincoln created a new norm.

While most children received Salk's vaccine in the months after its approval, the rate of vaccination among teenagers, who were also vulnerable to polio, was low. (Because polio was also known as "infantile paralysis," there was a misconception that only infants and toddlers could be stricken.) In 1956, before going on *The Ed Sullivan Show* to support the National Foundation for Infantile Paralysis (now known as the March of Dimes), Elvis Presley was photographed receiving his polio vaccination. The photographs were then published in newspapers across the country. That moment has been cited for a parabolic increase in vaccinations—a widely circulated if dubious statistic claimed the event raised immunization levels in the United States "from 0.6% to 80% in just 6 months!" Which might suggest that Elvis eradicated polio in America.

•

When I was young, people smoked cigarettes on airplanes. It is so unthinkable now, I had to check to make sure I was remembering correctly. How do we regard the prevalence of smoking in our near past, a norm in which nearly every demographic—including children and pregnant women—participated? Probably the same way people from environmentally conscientious countries regard Americans. The way our descendants will regard us.

Over the past decades, smoking norms changed: how many people smoke, how often, and where. What once was acceptable and even attractive became taboo, or at least unpleasant. So-called sin taxes and legislation helped—and resistance from industry lobbyists hurt—but the changes were primarily engendered by grassroots campaigns. Most people want to do what's good for the world, when it doesn't come at personal expense. Smoking is a physically addictive habit whose global implications (secondhand smoke and the health-care burden of cancer) feel remote. Yet the smoking rate in America has halved in my lifetime, largely because of grassroots campaigns. That sounds like a triumph, but it is a failure.

Why has smoking only halved? And what has taken so long? As early as 1949, 60 percent of Americans said that cigarette smoking was harmful to health. Information wasn't the obstacle then, and it certainly isn't the obstacle now. How do we reconcile the broadly accepted knowledge that smoking kills and the reality that there are still more smokers in America (nearly thirty-eight million) than there are people in Canada? Why would someone as aware and deliberate as Barack Obama still occasionally indulge a habit that, on average, reduces life span by twenty years? Probably for the same reason someone as aware and deliberate as Obama did not adequately address climate change. Many forces are stronger than a conceptual threat.

The tobacco industry has genetically altered cigarettes to be twice as addictive as they were fifty years ago and has disproportionately marketed them in low-income neighborhoods, often near schools. The industry has offered free cigarettes in housing projects and handed out tobacco coupons with food stamps. Despite the rising cost of cigarettes, nearly three in four smokers are from low-income neighborhoods.

Just as social movements like polio vaccination, #MeToo, smoking cessation, and environmentalism are advanced by concurrent forces, they are also hindered by concurrent forces.

•

Elvis's public vaccination might have contributed to the dramatic leap in immunization, but it didn't cause it. According to the historian Stephen Mawdsley,

> It was obviously a help in getting teenagers to take up the vaccine, but—intriguingly—not an overwhelming one. The real game-changer came through the teenagers themselves. With the help of the National Foundation for Infantile Paralysis, they established a group called Teens Against Polio, canvassed door-to-door, and set up dances where only vaccinated individuals could get in. It showed, almost for the first time, the power of teens in understanding and connecting with their own demographic.

Social change, much like climate change, is caused by multiple chain reactions that occur simultaneously. Both cause, and are caused by, feedback loops. No single factor can be credited for a hurricane, drought, or wildfire, just as no single factor can be credited for a decline in cigarette smoking—and yet in all cases, every factor is significant. When a radical change is needed, many argue that it is impossible for individual actions to incite it, so it's futile for anyone to try. This is exactly the opposite of the truth: the impotence of individual action is a reason for everyone to try.

On November 1, 2018, an estimated twenty thousand Google employees participated in a wave of international walkouts, mainly protesting the company's handling of sexual misconduct cases. The walkouts came together in less than a week, and more than 60 percent of Google's offices around the world participated. The collective response was especially significant because it challenged the kind of individualism that reigns as the dominating ethos of Silicon Valley. In a news release, the organizers of the protest said, "This is part of a growing movement, not just in tech, but across the country, including teachers, fast-food workers and others who are using their strength in numbers to make real change." A week later, Google granted the organizers' first request: it ended forced arbitration for sexual harassment. (Forced arbitration had previously prevented sexual harassment complaints from reaching the courts.) Days later, Facebook, Airbnb, and eBay followed suit.

In less than a week, an international protest was organized. A week later, Google changed its corporate policy. Days after that, three other major companies changed theirs. All this occurred in less than a month.

Polio couldn't have been cured without someone inventing a vaccine—that required an architecture of support (funding from the March of Dimes) and knowledge (Jonas Salk's medical breakthrough). But that vaccine couldn't have been approved without a wave of polio pioneers volunteering for a trial—their feelings were irrelevant; it was their participation in the collective action that allowed the cure to be brought to the public. And that approved vaccine would have been worthless if it had not become a social contagion, and therefore a norm—its success was the result of both top-down publicity campaigns and grassroots advocacy.

Who cured polio?

No one did.

Everyone did.

Open Your Eyes

Like its author, most people reading this book are not scientists on the level of Jonas Salk or celebrities on the level of Elvis. We live our lives without making ripples, much less waves. And when it comes to the planetary crisis, most of us feel lost inside the causes and effects, confused by the ever-changing statistics, frustrated by the rhetoric. We feel powerless, yet inexplicably calm. How are we, ordinary civilians, supposed to *do* anything about a crisis that we know about but don't believe in, that we have a muddled (at best) understanding of, and that we have no obvious ways to combat?

Watching Al Gore's *An Inconvenient Truth* was an intellectual and emotional revelation for me. When the screen went dark after the final image, our situation seemed perfectly clear, as did my responsibility to participate in the struggle. Like the tens of thousands of Americans who went straight to their local recruiting offices upon hearing the news of Pearl Harbor, I felt eager to enlist.

And when that film's credits rolled, at the moment of greatest enthusiasm to do whatever was asked to work against the imminent apocalypse that Gore had just delineated for us, suggested actions appeared on the screen. "Are you ready to change the way you live? The climate crisis can be solved. Here's how to start."

> Tell your parents not to ruin the world that you will live in.
> If you are a parent, join with your children to save the world they will live in.
> Switch to renewable sources of energy.
> Call your power company to see if they offer green energy. If they don't, ask them why not.
> Vote for leaders who pledge to solve this crisis.
> Write to congress. If they don't listen, run for congress.
> Plant trees, lots of trees.
> Speak up in your community.
> Call radio shows and write newspapers.
> Insist that America freeze CO_2 emissions.
> Join international efforts to stop global warming.
> Reduce our dependence on foreign oil; help farmers grow alcohol fuels.
> Raise fuel economy standards; require lower emissions from automobiles.
> If you believe in prayer, pray that people will find the strength to change.
> In the words of the old African proverb, when you pray, move your feet.
> Encourage everyone you know to see this movie.

I found that list frustratingly vague (*Call radio shows and say what exactly, and toward what end?*), unproductive (*I can tell my parents not to ruin the world that I will live in, and they can tell their parents the same, but at some point, doesn't someone have to actually do something?*), plainly unrealistic (*"Hello, Mr. President, it's me. Sorry I had you on hold—I was just helping some farmers grow alcohol fuels—but now that I have you, I insist that America freeze CO_2 emissions"*), and tautological in a way that would have been laughable if I weren't on the verge of tears (*Watch this movie so that you can encourage others to watch this movie so that they can encourage others to watch this movie*).

It is good to speak up, good to recycle, to plant trees, lots of trees. Those activities are good in the way that combing the skies for enemy planes that will never be there is good: to remind us that a war is being fought, to generate solidarity and will. According to a 2017 analysis, recycling and tree planting are among the most often recommended personal choices to combat climate change, but they aren't "high impact"—they are feelings more than actions. Among other actions that are considered to be important but aren't high impact: installing solar panels, conserving energy, eating locally, composting, washing clothes with cold water and hang-drying them, being sensitive to the amounts and kinds of packaging, buying organic food, replacing a conventional car with a hybrid. People who make those efforts—and *only* those efforts—are saying the word "fist" to an object they want to punch. Planes patrolling Midwestern skies without boots on the ground in Europe would have been suicidal.

There is a glaring absence in Gore's list, and its invisibility recurs in 2017's *An Inconvenient Sequel: Truth to Power*, with one minuscule exception. It is impossible to explain this omission as accidental without also accusing Gore of a kind of radical ignorance or malpractice. In terms of the scale of the error, it would be equivalent to a doctor prescribing physical exercise to a patient recovering from a heart attack without also telling him he needs to quit smoking, reduce his stress, and stop eating burgers and fries twice a day.

So why would Gore deliberately choose to leave it out? Almost certainly for fear that it would be distractingly controversial and dim the enthusiasm he had just worked so hard to ignite. It has also been largely absent from the websites of leading environmental advocacy organizations—although this now seems to be changing. It's also unmentioned in the celebrated book *Dire Predictions*, written by the climate scientists Michael E. Mann and Lee R. Kump to educate citizens on the fifth assessment report of the Intergovernmental Panel on Climate Change, released in 2014. After forecasting existential climate disasters, the authors recommend that we substitute clotheslines for electric dryers and commute by bicycle. Among their suggestions, there is no reference to the daily act that is, according to the research director of Project Drawdown—a collection of nearly two hundred environmental scientists and thought leaders dedicated to identifying and modeling substantive solutions to address climate change—"the most important contribution every individual can make to reversing global warming."

In America, environmentalists have been fighting an uphill battle from the start, facing the challenges of educating citizens about something abstract and hard to believe, as well as enormous resistance from the fossil fuel industry and, after a brief period of bipartisan cooperation, from most of a political party. If they have spent decades trying to persuade the public that pulling carbon out of the earth and burning it causes climate change, and people *still* elect a president who has called global warming a Chinese hoax, how can they expect to launch a conversation that challenges fundamental aspects of our personal, familial, and cultural identities? Some organizations and public figures fear losing the momentum and support they worked so hard to achieve. Some fear being accused of hypocrisy. Some fear that shifting the attention away from fossil fuels would undermine decades of efforts to fight the global superpower of Big Oil.

The politics and psychology of activism matter. Every argument is essentially a story, and certain stories (Rosa Parks) work better than others (Claudette Colvin). Sometimes it is best to hide a challenging reality in service of ultimately leading people back to it. But how truthful is an inconvenient truth that omits one of the greatest contributors to our planetary crisis, which also happens to be the easiest to correct? What if winning the most important war we will ever fight—the struggle for our way of life, and for life itself—depended on a collective act that, relative to the scale of our war, is proportionate to turning off the lights at night? Shouldn't we at least talk about it? Even for believers, isn't praying that people will find the strength to change incomplete until we reveal the changes in question?

Our ways of addressing the planetary crisis aren't working. Al Gore deserves his Nobel Prize, but the change he has inspired isn't nearly enough—a fact he readily admits in *An Inconvenient Sequel*. The institutions of environmentalism deserve our support, but their accomplishments aren't close to sufficient, either. Anyone who knows the science, and is willing to acknowledge the most inconvenient truth of all, will agree that we are doing far too little, far too slowly, and that our present course leads to our own destruction.

According to one estimate, electricity use accounts for

25 percent of annual greenhouse gas emissions. Agriculture accounts for 24 percent, mostly from animal agriculture. Manufacturing also accounts for 24 percent. Transportation: 14 percent. Buildings: 6 percent. Miscellaneous sources account for the remainder. All these emissions need to fall to zero, which will require innovation and cooperation—a feat that will be impossible if we don't start talking about every contributing sector.

The Paris accord's goal of keeping global warming below 2 degrees Celsius (3.6 degrees Fahrenheit), thought of as an ambitious target, is the outside edge of cataclysm. Even if we are miraculously able to achieve it—recent statistical models put the likelihood at 5 percent—we will be living in a far less hospitable world than the one we know, and many of the changes set in motion would be at best irreversible and at worst self-amplifying. If we defy the heavily stacked odds and limit global warming to 2 degrees:

- Sea levels will rise by 1.6 feet, flooding coastlines across the globe. Dhaka (population 18 million), Karachi (15 million), New York (8.5 million), and dozens of other metropolises will be effectively uninhabitable; 143 million people are projected to become climate migrants.
- Armed conflict will increase by an estimated 40 percent because of climate change.
- Greenland will tip into irreversible melt.
- Between 20 and 40 percent of the Amazon will be destroyed.
- The European heat wave of 2003—which cost more than seventy thousand lives and thirteen billion euros

in crop losses, and brought the Rivers Po, Rhine, and Loire to historic lows—will be the annual norm.

- Human mortality will dramatically increase because of heat waves, floods, and droughts. There will be rampant increases in asthma and other respiratory illnesses. The number of people at risk of malaria will increase by several hundred million.
- Four hundred million people will suffer from water scarcity.
- Warmer oceans will irreparably damage 99 percent of coral reefs, disrupting ecosystems for nine million species.
- Half of all animal species will face extinction.
- A total of 60 percent of all plant species will face extinction.
- Wheat yields will be reduced by 12 percent, rice by 6.4 percent, maize by 17.8 percent, and soybeans by 6.2 percent.
- Global GDP per capita will drop by an estimated 13 percent.

These are some upsetting statistics, whose emotional impact is unlikely to survive to the end of this sentence. That is, the horrific future they describe will be acknowledged by most readers of this book and believed by few. I am sharing these figures with the hope that you will believe them. But I don't believe them.

Meeting the goals of the Paris accord and living in the world described above is the *best-case scenario*. The few experts who think we have any realistic chance of achieving those goals are either deceiving themselves or, more likely, weaponizing optimism to alter the odds. The truth is, even if we were to somehow turn off all lights and outlaw all automobiles, without making the change that those like Gore know but don't talk about, we have *no chance*.

When I was a boy, my father told me that the best way to get rid of a bee was not to run from it, swat at it, or even stand motionless, but to close my eyes and count to ten. "It will work every single time," he said. "And if it doesn't, count to twenty." It did work, but advice that works isn't always good advice.

There were a number of subjects my family didn't talk about when I was young, the echoing traumas of the Holocaust chief among them. Who could blame us for closing our eyes until that threat seemed to go away? I have my own family now, and my own avoided subjects. I don't blame myself for wanting to protect my children (and myself) from pain. Those acts of willful blindness are acts of love. But I will have to blame myself if closing my eyes permits far greater pain to grow, in the same way that I will have to blame myself if one day I am diagnosed with an illness that would have been treatable had I gone to a doctor before symptoms expressed themselves. I think of myself as health conscious, but I haven't had a physical in years. Like you, I think of myself as many things, as if the thinking made it so. In the meantime, while I think—while you think, while we think—our actions and inactions create and destroy the world.

•

Imagine the scene: More than 150,000 soldiers are storming the beaches at Normandy. It is the largest amphibious invasion ever mounted. Even at the time, it is recognized as a hinge moment in history. The operation is happening now, June 6, 1944, because the full moon is necessary for the tide and for illumination. The Allied planning for the invasion has involved creating more than seventeen million maps, training four thousand new cooks to feed the collected men, constructing a duplicate of Nazi shore defenses for training, and sewing hundreds of dummies—sometimes dressed with boots and helmets, sometimes equipped with recordings of gunfire and explosions—to be dropped at various locations to divide German attention. The soldiers wading onto the beach have come from a dozen countries. They are supposed to be no younger than eighteen and no older than forty-one, although younger and older men have enlisted with falsified documents. The landing crafts push forward, releasing as many as two hundred men at a time into the storm of war.

A child's father pulls the trigger of his rifle, hears the crack of the shot. He is unaware that he just fired a blank.

A Jewish soldier from Pittsburg fires ten blanks per second from an M1919 machine gun.

Someone's piano teacher's hand is shaking too violently to fire the first shot from a pistol loaded with blanks.

Someone's favorite outfielder throws a grenade as deadly as a baseball.

The bayonet at the end of someone's child's rifle ends in a blunt stump.

Because of the chaos of the battlefield, and because each soldier's experience is wholly consuming for him, and because it *feels* like fighting, no one realizes that it *only* feels like fighting—that he is as effective a soldier as the dummies parachuting from the sky.

•

Close your eyes and count to ten.

Advice that seems to work doesn't always work.

The final time I closed my eyes to get rid of a bee, the bee stung me on my eyelid. My eye swelled and wouldn't open. As if that bee's father had told it the best way to get rid of a human was to land on its closed eye.

Ours Alone

General Eisenhower prepared a statement in the event that the D-Day invasion was repelled:

> Our landings in the Cherbourg-Havre area have failed to gain a satisfactory foothold and I have withdrawn the troops. My decision to attack at this time and place was based upon the best information available. The troops, the air and the Navy did all that bravery and devotion to duty could do. If any blame or fault attaches to the attempt it is mine alone.

Of his historic moon walk, Neil Armstrong said:

> When you have hundreds of thousands of people all doing their job a little better than they have to, you get an improvement in performance. And that's the only reason we could have pulled this whole thing off.

Show Your Hands

This is a book about the impacts of animal agriculture on the environment. Yet I have managed to conceal that for the previous sixty-three pages. I have navigated away from the subject for the same reasons that Gore and others have: fear that it is a losing hand. I evaded even while I was critiquing Gore for his evasion—I never mentioned what he never mentions. I felt sure, as Gore must have, that it was the right strategy. Conversations about meat, dairy, and eggs make people defensive. They make people annoyed. No one who isn't a vegan is eager to go there, and the eagerness of vegans can be a further turnoff. But we have no hope of tackling climate change if we can't speak honestly about what is causing it, as well as our potential, and our limits, to change in response. Sometimes a fist needs the word "fist" written across it, so I'll name it now: we cannot save the planet unless we significantly reduce our consumption of animal products.

This book is an argument for a collective act to eat differently—specifically, no animal products before dinner. That is a difficult argument to make, both because the topic is so fraught and because of the sacrifice involved. Most people like the smell and taste of meat, dairy, and eggs. Most people value the roles animal products play in their lives and aren't prepared to adopt new eating identities. Most people have eaten animal products at almost every meal since they were children, and it's hard to change lifelong habits, even when they aren't freighted with pleasure and identity. Those are meaningful challenges, not only worth acknowledging but necessary to acknowledge. Changing the way we eat is simple compared with converting the world's power grid, or overcoming the influence of powerful lobbyists to pass carbon-tax legislation, or ratifying a significant international treaty on greenhouse gas emissions—but it isn't simple.

In my early thirties, I spent three years researching factory farming and wrote a book-length rejection of it called *Eating Animals*. I then spent nearly two years giving hundreds of readings, lectures, and interviews on the subject, making the case that factory-farmed meat should not be eaten. So it would be far easier for me not to mention that in difficult periods over the past couple of years—while going through some painful personal passages, while traveling the country to promote a novel when I was least suited for self-promotion—I ate meat a number of times. Usually burgers. Often at airports. Which is to say, meat from precisely the kinds of farms I argued most strongly against. And my reason for doing so makes my hypocrisy even more pathetic: they brought me comfort. I can imagine this confession eliciting some ironic comments and eye-rolling, and some giddy accusations of fraudulence. Other readers may find it genuinely disturbing—I wrote at length, and passionately, about how factory farming tortures animals and destroys the environment. How could I argue for radical change, how could I raise my children as vegetarians, while eating meat *for comfort*?

I wish I had found comfort elsewhere—in something that would have provided it in a lasting way and that wasn't anathema to my convictions—but I am who I am, and I did what I did. Even while I was working on this book, and having my commitment to vegetarianism—which had been driven by the issue of animal welfare—deepened by a full awareness of meat's environmental toll, rarely a day has passed when I haven't craved it. At times I've wondered if my strengthening intellectual rejection of it has fueled a strengthening desire to consume it. Whatever the case, I've had to come to terms with the fact that while actions might be at least somewhat responsive to will, cravings aren't. I have felt a version of Felix Frankfurter's knowledge-without-belief, and that has led me to some real struggling, and at times to extreme hypocrisy. I find it almost unbearably embarrassing to share this. But it needs to be shared.

While I was promoting *Eating Animals*, people frequently asked me why I wasn't vegan. The animal welfare and environmental arguments against dairy and eggs are the same as those against meat, and often stronger. Sometimes I would hide behind the challenges of cooking for two finicky children. Sometimes I would bend the truth and describe myself as "effectively vegan." In fact, I had no answer, other than the one that felt too shameful to voice: my desire to eat cheese and eggs was stronger than my commitment to preventing cruelty to animals and the destruction of the environment. I found some relief from that tension by telling other people to do what I couldn't do myself.

Confronting my hypocrisy has reminded me how difficult it is to live—even to try to live—with open eyes. Knowing that it will be tough helps make the efforts possible. *Efforts*, not effort. I cannot imagine a future in which I decide to become a meat-eater again, but I cannot imagine a future in which I don't want to eat meat. Eating consciously will be one of the struggles that span and define my life. I understand that struggle not as an expression of my uncertainty about the right way to eat, but as a function of the complexity of eating.

We do not simply feed our bellies, and we do not simply modify our appetites in response to principles. We eat to satisfy primitive cravings, to forge and express ourselves, to realize community. We eat with our mouths and stomachs, but also with our minds and hearts. All my different identities—father, son, American, New Yorker, progressive, Jew, writer, environmentalist, traveler, hedonist—are present when I eat, and so is my history. When I first chose to become vegetarian, as a nine-year-old, my motivation was simple: do not hurt animals. Over the years, my motivations changed—because the available information changed, but more importantly, because my life changed. As I imagine is the case for most people, aging has proliferated my identities. Time softens ethical binaries and fosters a greater appreciation of what might be called the messiness of life.

If I'd read the previous sentences in high school, I'd have dismissed them as a bursting sack of self-serving bullshit—*messiness of life?*—and been deeply disappointed by the flimsy person I was to become. I'm glad that I was who I was then, and I hope that other young people have the same inflexible idealism. But I'm glad that I am who I am now, not because it is easier but because it is in better dialogue with my world, which is different from the world I occupied twenty-five years ago.

There is a place at which one's personal business and the business of being one of seven billion earthlings intersect. And for perhaps the first moment in history, the expression "one's time" makes little sense. Climate change is not a jigsaw puzzle on the coffee table, which can be returned to when the schedule allows and the feeling inspires. It is a house on fire. The longer we fail to take care of it, the harder it becomes to take care of, and because of positive feedback loops—white ice melting to dark water that absorbs more heat; thawing permafrost releasing huge amounts of methane, one of the worst greenhouse gases—we will very soon reach a tipping point of "runaway climate change," when we will be unable to save ourselves, no matter our efforts.

We do not have the luxury of living in our time. We cannot go about our lives as if they were only ours. In a way that was not true for our ancestors, the lives we live will create a future that cannot be undone. Imagine if history were such that if Lincoln hadn't abolished slavery in 1863, then America would be condemned to uphold the institution of slavery for the rest of time. Imagine if the right of two people of the same sex to marry depended entirely and eternally on Obama's conversion in 2012. When speaking about moral progress, Obama often quoted Martin Luther King's statement that "the arc of the moral universe is long, but it bends toward justice." In this unprecedented moment, the arc could irreparably snap.

There are several pivotal moments in the Bible when God asks people where they are. The two most cited instances are when he finds Adam hiding after eating the forbidden fruit and says "Where are you?," and when he calls to Abraham before asking him to sacrifice his only son. Clearly an omniscient God knows where his creations are. His questions are not about the location of a body in space but about the location of a self within a person.

We have our own modern version of this. When we think back on moments when history seemed to happen before our eyes—Pearl Harbor, the assassination of John F. Kennedy, the fall of the Berlin Wall, September 11—our reflex is to ask others where they were when it happened. Yet as with God in the Bible, we are not really trying to establish someone's coordinates. We are asking something deeper about their connection to the moment, with the hope of situating our own.

The word "crisis" derives from the Greek *krisis*, meaning "decision."

The environmental crisis, though a universal experience, doesn't feel like an event that we are a part of. It doesn't feel like an event at all. And despite the trauma of a hurricane, wildfire, famine, or extinction, it's unlikely that a weather event will inspire a "Where were you when . . ." question of anyone who didn't live through it—perhaps not even of those who did live through it. It's all just weather. Just environmental.

But future generations will almost certainly look back and wonder where we were in the biblical sense: Where was our selfhood? What decisions did the crisis inspire? Why on earth—why on *Earth*—did we choose our suicide and their sacrifice?

Perhaps we could plead that the decision wasn't ours to make: as much as we cared, there was nothing we could do. We didn't know enough at the time. Being mere individuals, we didn't have the means to enact consequential change. We didn't run the oil companies. We weren't making government policy. Perhaps we could argue, as Roy Scranton does in his *New York Times* essay "Raising My Child in a Doomed World," that "we [were] not free to choose how we live[d] any more than we [were] free to break the laws of physics." The ability to save ourselves, and save them, was not in our hands.

But that would be a lie.

•

While information is not sufficient—without belief, knowing is *only knowing*—it is necessary for making a good decision.

Awareness of Nazi atrocities didn't shake Felix Frankfurter's conscience, but without that awareness, he would have no reason to be asked, or to ask himself, "Where are you?" Knowing is the difference between a grave error and an unforgivable crime.

With respect to climate change, we have been relying on dangerously incorrect information. Our attention has been fixed on fossil fuels, which has given us an incomplete picture of the planetary crisis and led us to feel that we are hurling rocks at a Goliath far out of reach. Even if they are not persuasive enough on their own to change our behavior, facts can change our minds, and that's where we need to begin. We know we have to do something, but *we have to do something* is usually an expression of incapacitation, or at least uncertainty. Without identifying the thing that we have to do, we cannot decide to do it.

The next section of this book will correct the picture by explaining the connection between animal agriculture and climate change. I have condensed what could have been several hundred pages of prose into a handful of the most essential facts. And I have not included important complementary narratives—the other kinds of destruction factory farming wreaks on the environment, like water pollution, ocean dead zones, and loss of biodiversity; the cruelty that is fundamental to contemporary animal agriculture; the health and societal effects of eating unprecedentedly large amounts of meat, dairy, and eggs. This book is not a comprehensive explanation of climate change, and it is not a categorical case against eating animal products. It is an exploration of a decision that our planetary crisis requires us to make.

The word "decision" derives from the Latin *decidere*, which means "to cut off." When we decide to turn off the lights during a war, refuse to move to the back of the bus, flee our shtetl with our sister's shoes, lift a car off a trapped person, make way for an ambulance, drive home through the night from Detroit, rise for a wave, take a selfie, participate in a medical trial, attend a Thanksgiving meal, plant a tree, wait in line to vote, or eat a meal that reflects our values, we are also deciding to cut off the possible worlds in which we don't do those things. Every decision requires loss, not only of what we might have done otherwise but of the world to which our alternative action would have contributed. Often that loss feels too small to notice; sometimes it feels too large to bear. Usually, we just don't think about our decisions in those terms. We live in a culture of historically unprecedented acquisition, which so often asks us and enables us to attain. We are prompted to define ourselves by what we have: possessions, dollars, views and likes. But we are revealed by what we release.

Climate change is the greatest crisis humankind has ever faced, and it is a crisis that will always be simultaneously addressed together and faced alone. We cannot keep the kinds of meals we have known and also keep the planet we have known. We must either let some eating habits go or let the planet go. It is that straightforward, that fraught.

Where were you when you made your decision?

II. HOW TO PREVENT THE GREATEST DYING

Degrees of Change

- From one hundred thousand to ten thousand years ago, mastodons, mammoths, dire wolves, saber-toothed cats, and giant beavers roamed a world of ice. The average global temperature was four to seven degrees Celsius colder than it is today.

- Fifty million years ago, the Arctic was filled with tropical rainforests. Crocodiles, turtles, and alligators lived in the polar forests of what is now Canada and Greenland. Two-hundred-pound penguins waddled in Australia, and palm trees grew in Alaska. There were no polar ice caps. The Antarctic seas were warm enough for a balmy swim, and around the equator, the oceans were the temperature of a hot tub. Earth was five to eight degrees Celsius warmer than it is today.

- As with body temperature, a few degrees can be the difference between health and crisis.

The First Crisis

- There have been five mass extinctions. All but the one that killed the dinosaurs were caused by climate change.

- The most lethal mass extinction occurred 250 million years ago, when volcanic eruptions released enough carbon dioxide (CO_2) to warm the oceans by about ten degrees Celsius, ending 96 percent of marine life and 70 percent of life on land. The event is known as the Great Dying.

- Many scientists call the geological age from the Industrial Revolution to the present the Anthropocene, the period during which human activity has been the dominant influence on the earth.

- We are now experiencing the sixth mass extinction, often referred to as the Anthropocene extinction.

- Taking into account natural mechanisms that influence climate, human activity is responsible for 100 percent of the global warming that has occurred since the beginning of the Industrial Revolution, around 1750.

- The current climate change is the first caused by an animal and not by a natural event.

- The sixth mass extinction is the first climate crisis.

The First Farming

- If human history were a day, we were hunter-gatherers until about ten minutes before midnight.
- Humans represent 0.01 percent of life on Earth.
- Since the advent of agriculture, approximately twelve thousand years ago, humans have destroyed 83 percent of all wild mammals and half of all plants.

Our Planet Is an Animal Farm

- Globally, humans use 59 percent of all the land capable of growing crops to grow food for livestock.

- One-third of all the fresh water that humans use goes to livestock, while only about one-thirtieth is used in homes.

- Seventy percent of the antibiotics produced globally are used for livestock, weakening the effectiveness of antibiotics to treat human diseases.

- Sixty percent of all mammals on Earth are animals raised for food.

- There are approximately thirty farmed animals for every human on the planet.

Our Population Growth Is Radical

- Before the Industrial Revolution, the average life expectancy in Europe was about thirty-five years. It is now about eighty.

- It took two hundred thousand years for the human population to reach one billion, but only two hundred more years to reach seven billion.

- Every day, 360,000 people—roughly equal to the population of Florence, Italy—are born.

Our Animal Farming Is Radical

- In 1820, 72 percent of the American workforce was directly involved in agriculture. Today, 1.5 percent is.

- Like the video game console, the factory farm was an invention of the 1960s. Before then, food animals were raised outdoors in sustainable concentrations.

- Between 1950 and 1970, the number of American farms declined by half, the number of people employed in farming declined by half, and the size of the average farm doubled. During that time, the size of the average chicken also doubled.

- In 1966, distorting contact lenses were invented to make it harder for chickens to see their increasingly unnatural surroundings, thereby easing the stress that caused violent pecking and cannibalism. The lenses were considered too burdensome for farmers, so automated debeakers—which burn off the ends of chickens' faces—became the industry norm.

- In 2018, more than 99 percent of the animals eaten in America were raised on factory farms.

Our Eating Is Radical

- The current level of meat and dairy consumption is the equivalent of every person alive on the planet in 1700 eating 950 pounds of meat and drinking 1,200 gallons of milk every day.

- There are twenty-three billion chickens living on Earth at any given time. Their combined mass is greater than that of all other birds on our planet. Humans eat sixty-five billion chickens per year.

- On average, Americans consume twice the recommended intake of protein.

- People who eat diets high in animal protein are four times as likely to die of cancer as those who eat diets low in animal protein are.

- Smokers are three times as likely to die of cancer as nonsmokers are.

- In America, one out of every five meals is eaten in a car.

Our Climate Change Is Radical

- We are currently in the Quaternary glaciation, a period with continental and polar ice sheets. Such a period is more commonly known as an ice age.

- According to models of cyclical climate change, Earth should be experiencing a period of slight cooling right now.

- Nine of the ten warmest years on record have occurred since the first YouTube video, "Me at the Zoo," was posted, in 2005.

- During the Great Dying, a series of Siberian volcanoes produced enough lava to cover the United States up to three Eiffel Towers deep.

- Humans are now adding greenhouse gases to the atmosphere ten times faster than the volcanoes did during the Great Dying.

Why Greenhouse Gases Matter

- Sunlight passes through the atmosphere and heats the Earth. A portion of that heat bounces back into space. Greenhouse gases in the atmosphere trap some of the outgoing heat, as a blanket traps body heat.

- Life on Earth depends on the greenhouse effect. Without it, Earth's average temperature would be near zero degrees Fahrenheit, instead of fifty-nine degrees.

- CO_2 accounts for 82 percent of the greenhouse gases emitted by human activity. The majority is emitted by industry, transport, and electrical use.

- For the eight hundred thousand years before the Industrial Revolution, concentrations of greenhouse gases in our atmosphere remained stable. Since the Industrial Revolution, the concentration of CO_2 in the atmosphere has increased by about 40 percent.

- Methane and nitrous oxide are the second and third most prevalent greenhouse gases in the atmosphere. Animal agriculture is responsible for 37 percent of anthropogenic methane emissions and 65 percent of anthropogenic nitrous oxide emissions.

- Between the advent of factory farming in the 1960s and 1999, concentrations of nitrous oxide in the atmosphere grew about two times faster, and concentrations of methane grew six times faster, than they had over any previous forty-year period during the last two thousand years.

Climate Change Is a Ticking Time Bomb

- Different climate scientists have given different deadlines by which we must halt greenhouse gas emissions (GHGs). Such statements usually take the form "We have X years to solve climate change."

- Climate change is not a disease that can be managed, like diabetes; it is an event like a cancerous tumor that needs to be removed before the cells fatally multiply. The planet can handle only so much warming before positive feedback loops create "runaway climate change."

- One of the most powerful feedback loops is called the albedo effect. White ice sheets reflect sunlight back into the atmosphere. Dark oceans absorb sunlight. As the planet warms, there is less ice to reflect sunlight, and more dark ocean and land to absorb it. Oceans become hotter, melting ice faster.

- The former United Nations climate chief Christiana Figueres has said that we have until 2020 to avoid temperature thresholds leading to runaway, irreversible climate change.

Because Climate Change Is a Ticking Time Bomb, Not All Greenhouse Gases Matter Equally

- Methane has 34 times the global warming potential (GWP)—the ability to trap heat—as CO_2 does over a century. Over two decades, methane is 86 times as powerful. If CO_2 were the thickness of an average blanket, imagine methane as a blanket thicker than LeBron James is tall.

- Nitrous oxide has 310 times the GWP of CO_2. Imagine a blanket so thick you could commit suicide by jumping off it.

- When global emissions are calculated, greenhouse gases are converted to carbon dioxide equivalents (CO_2e). Calculations are usually based on a hundred-year timescale. This means that one metric ton of methane should be counted as thirty-four metric tons of CO_2 in an overall greenhouse gas assessment.

- We can think of our atmosphere as a budget and our emissions as expenses: because methane and nitrous oxide are significantly larger greenhouse expenses than CO_2 in the short term, they are the most urgent to cut. Because they are primarily created by our food choices, they are also easier to cut.

Why Deforestation Matters

- Trees are "carbon sinks," which means they absorb CO_2.

- Imagine a bathtub filling up with water. If the drain slows, the tub will fill up more quickly. This is similar to the earth's photosynthetic capacity: already, humans are pumping greenhouse gases into the atmosphere at a rate that exceeds Earth's ability to regulate them, but vegetation currently stores a substantial amount of CO_2—about one-quarter of anthropogenic emissions, or about half a century's worth of emissions at the current rate.

- The more forests we destroy, the closer we come to plugging the drain.

- Allowing tropical land currently used for livestock to revert to forest could mitigate more than half of all anthropogenic GHGs.

- Trees are 50 percent carbon. Like coal, they release their stores of CO_2 when burned.

- Forests contain more carbon than do all exploitable fossil-fuel reserves.

- The cutting and burning of forests is responsible for at least 15 percent of global GHGs per year. According to *Scientific American*, "By most accounts, deforestation in tropical rainforests adds more carbon dioxide to the atmosphere than the sum total of cars and trucks on the world's roads."

- About 80 percent of deforestation occurs to clear land for crops for livestock and grazing.

- Every year, wildfires in California create more greenhouse gas emissions than the state's progressive environmental policies save.

- Burning forests is like further opening the tap while clogging the drain.

Not All Deforestation Matters Equally

- In 2018, Brazil elected Jair Bolsonaro as president.
- Bolsonaro campaigned on a plan to develop previously protected swaths of the Amazon (i.e., deforestation).
- It has been estimated that Bolsonaro's policy would release 13.2 gigatons of carbon—more than two times the annual emissions of the entire United States.
- Animal agriculture is responsible for 91 percent of Amazonian deforestation.

Animal Agriculture Causes Climate Change

- As they digest food, cattle, goats, and sheep produce a significant amount of methane, which is mostly belched but also exhaled, farted, and passed in the waste of the animal.

- Livestock are the leading source of methane emissions.

- Nitrous oxide is emitted by livestock urine, manure, and the fertilizers used for growing feed crops.

- Livestock are the leading source of nitrous oxide emissions.

- Animal agriculture is the leading cause of deforestation.

- According to the United Nations Framework Convention on Climate Change, if cows were a country, they would rank third in greenhouse gas emissions, after China and the United States.

Animal Agriculture Is a/the Leading Cause of Climate Change

- When assessing animal agriculture's overall contribution to greenhouse gas emissions, estimates range dramatically depending on what is included in the calculation.

- The Food and Agriculture Organization (FAO) of the United Nations asserts that livestock are a leading cause of climate change, responsible for approximately 7,516 million tons of CO_2e emissions per year, or 14.5 percent of annual global emissions.

- The FAO calculation includes the CO_2 emitted when forests are cleared for animal feed crops and pastures, but it does not take into account the CO_2 that those forests can no longer absorb. (Imagine a life insurance policy that covered the cost of the funeral but not future lost wages.) Among other things not included in its calculation is the CO_2 exhaled by farmed animals, even though, in the words of one environmental-assessment specialist, "livestock (like automobiles) are a human invention and convenience, not part of pre-human times, and a molecule of CO_2 exhaled by livestock is no more natural than one from an auto tailpipe."

- When researchers at the Worldwatch Institute accounted for emissions that the FAO overlooked, they estimated that livestock are responsible for 32,564 million tons of CO_2e emissions per year, or 51 percent of annual global emissions—more than all cars, planes, buildings, power plants, and industry combined.*

- We do not know for sure if animal agriculture is *a* leading cause of climate change or *the* leading cause of climate change.

- We know for sure that we cannot address climate change without addressing animal agriculture.

*An explanation of these different calculations can be found in the appendix.

It Will Be Impossible to Defuse the Ticking Time Bomb Without Reducing Our Consumption of Animal Products

- Scientists estimate that to keep global warming at or below two degrees Celsius—the goal of the Paris accord—we have a CO_2e emissions budget of 565 gigatons by 2050.

- According to a recent Johns Hopkins University report on the role of diet in climate control, "If global trends in meat and dairy intake continue, global mean temperature rise will more than likely exceed 2° C, even with dramatic emissions reductions across non-agricultural sectors."

- Home-front efforts during WWII were not enough, on their own, to win the war, but the war could not have been won without home-front efforts. Changing how we eat will not be enough, on its own, to save the planet, but we cannot save the planet without changing how we eat.

Not All Actions Are Equal

- The most optimistic estimates suggest that, even assuming international cooperation, a global conversion to wind, water, and solar power would take more than twenty years and require a hundred-trillion-dollar investment.

- Hans Joachim Schellnhuber, director of the Potsdam Institute for Climate Impact Research: "The maths is brutally clear: while the world can't be healed within the next few years, it may be fatally wounded by negligence [before] 2020."

- Adjusted for inflation, the global cost of WWII was fourteen trillion dollars.

- The four highest-impact things an individual can do to tackle climate change are eat a plant-based diet, avoid air travel, live car-free, and have fewer children.

- Of those four actions, only plant-based eating immediately addresses methane and nitrous oxide, the most urgently important greenhouse gases.

- Most people are not in the process of deciding whether to have a baby.

- Eighty-five percent of Americans drive to work. Few drivers can simply decide to stop using their cars.

- For Americans, 29 percent of air travel in 2017 was for business purposes, and 21 percent was for "personal non-leisure purposes." Businesses must rely more on remote communication, "personal non-leisure" flights must be reduced, and personal leisure flights can and must be cut, but the fact remains that a sizable portion of air travel is unavoidable.

- Everyone will eat a meal relatively soon and can immediately participate in the reversal of climate change.

Not All Foods Are Equal

- Pounds of CO_2e associated with a serving of each food:

  Beef: 6.61
  Cheese: 2.45
  Pork: 1.72
  Poultry: 1.26
  Eggs: 0.89
  Milk: 0.72
  Rice: 0.16
  Legumes: 0.11
  Carrots: 0.07
  Potatoes: 0.03

- Not eating animal products for breakfast and lunch has a smaller CO_2e footprint than the average full-time vegetarian diet.

How to Prevent the Greatest Dying

- To meet the Paris accord's two-degree goal, an individual's annual CO_2e budget must not exceed 2.1 metric tons by 2050.

- While citizens of different countries have dramatically different CO_2e footprints—the average American's is 19.8 metric tons per year, the average Frenchman's is 6.6 metric tons per year, and the average Bangladeshi's is 0.29 metric tons per year—the average global citizen has a CO_2e footprint of approximately 4.5 metric tons per year.

- Not eating animal products for breakfast and lunch saves 1.3 metric tons per year.

III. ONLY HOME

Mapping Our Vision

There came a point at which the inhabitants of Mars were no longer able to deny the warming of their planet or the scale of the destruction to come. In a last, desperate attempt to maintain their civilization, they dug vast canals connecting the poles of the planet to the expanse of scorched land that covered the rest of its surface. The annual melt of the polar ice caps would produce water to grow enough crops to sustain at least another generation.

This final struggle against extinction was documented by the astronomer Percival Lowell from his private observatory in Flagstaff, Arizona, at the end of the nineteenth century. Lowell was no quack—he was elected a fellow of the American Academy of Arts and Sciences and is credited for leading the effort that culminated in the discovery of Pluto—but given that Mars's "non-natural features" couldn't be observed by any other astronomers of his time, his theory, which enthralled the public, was rejected by the scientific community. He continued to observe and make meticulous drawings of the Martian canals, and continued to insist, until his death, in 1916, that they were the last heroic attempts of a dying civilization to save itself.

Lowell didn't initiate the search for Martian canals. In 1877, the Italian astronomer Giovanni Schiaparelli reported observing *canali* on Mars, thus launching a search among English-speaking astronomers for nonnatural features on the planet's surface. Lowell was the only one to confirm Schiaparelli's observations. Alas, the Italian word *canali* means "channels" (a naturally occurring feature, of which there are many on Mars), not "canals," and was mistranslated into English.

When a NASA Mariner spacecraft flew by Mars in 1965, capturing the first photographs of the planet's surface, the existence of canals was conclusively disproved. If Mars was once inhabited by intelligent life, either that civilization covered its tracks or evidence of its existence had been erased by time—as scientists say will be the case about twenty thousand years after the disappearance of humankind from Earth.

But it took another forty years to explain what Lowell had been observing and documenting all that time.

•

I am sitting at my grandmother's bedside as I type these words. She has been living with my parents for the last few years, after a stint in an assisted-living facility proved to be too stressful for her. At this point, she sleeps most of the day. My mother tells me that my grandmother's wish is to be woken up whenever someone comes to see her. It goes against so many of my instincts—never wake a sleeping baby, never wake a dying grandmother—but in this case, I act on what I know, not on what I feel. Her smile raises with her eyelids as if they are connected by threads.

She is as mentally present as ever. Despite—or because of—there being so many ultimate things to talk about, it feels as if there isn't anything to talk about. So most of the time, we just sit here quietly. Sometimes she stays awake, sometimes she falls back asleep. Sometimes I go downstairs and hang out with my parents while she rests. Sometimes, like now, I stay here. One of the ways that I've filled the hours has been to drive around the city to the neighborhoods and sites of my youth: Mr. L's restaurant is gone; Higgers Drugs is gone; Politics and Prose bookstore moved across the street and spread like an empire; the Sheridan School playground has been overwhelmed by new classrooms; Fort Reno is still there, although Fugazi is no longer a band.

Everything is the wrong size. The "big hill" that my brother and I used to dare each other to bike down without braking is at most a gentle slope. The walk to school, which I remember taking nearly an hour, is only six blocks. But the school itself, which I remember as small, is enormous—many times larger than the school my children now attend. My sense of scale isn't skewed in a particular direction, but it is badly skewed.

The strangest thing to reencounter was the home where I lived for the first nine years of life. In this case, it was not the physical scale that was skewed, but the emotional scale. I was sure I'd have strong feelings revisiting it for the first time in decades, but it was merely interesting, and I was happy enough to leave after ten minutes.

A few years ago, an artist conducted a series of lengthy interviews with each of my brothers and me, drawing out memories of our shared childhood home. *What color is the front door? What do you see upon entering? Is the floor bare or covered? Approximately how many stairs are there? What do the banisters look like? Do the windows have coverings of any kind? How many bulbs are in the light fixture?* (All her questions were in the present tense.) She then produced three distinct floor plans of the house, corresponding to our memories. The discrepancies were astounding: different configurations of rooms, different scales, even a different number of floors. How could that be? It was not some building we'd entered only a few times. It was the home in which we were raised. Maybe her experiment proved that memory is even less reliable than we suspect, or that we were too busy being kids to pay attention to our surroundings. But a far more unsettling possibility is that home—which we think of as being essential to the stories we invent and the stories we believe—isn't nearly as powerful as we assume. Maybe home, in the end, is just a place.

•

After the Roman Empire's fall, exotic plants bloomed across the Colosseum's bloodstained ground, plants found nowhere else in Europe. They overcame the balustrades, choked the columns, relentlessly grew and grew. For a time, the Colosseum was the world's greatest botanical garden, if an unintentional one. The seeds had been unknowingly transported in the pelts of the bulls, bears, tigers, and giraffes brought from thousands of miles away for the gladiators to slaughter. The plants occupied the Roman Empire's absence.

When my grandmother and I used to go on weekend walks through the park, she would take a moment of rest at every bench—it would probably be more accurate to describe those Sunday hours as weekend rests punctuated by moments of walking. Usually we would sit in silence. Sometimes she would give me life advice: "Marry someone a little bit less intelligent than you"; "It's just as easy to fall in love with a rich person"; "You paid for the bread in the basket, so you should take it." More than once she placed her enormous hand on my knee and told me, "You are my revenge."

This statement has always puzzled me, and I've arrived at different interpretations over the years. "Revenge" comes from the Latin word for "vengeance," *vindicare*, meaning "to set free" or "to lay claim to." To set something free again. To reclaim. The ultimate revenge against a genocide that is meant to eradicate you and your people is to create a family. The ultimate revenge against a force that tries to claim and imprison you is to set yourself free again, to reclaim your life. Maybe when she looked at her children and grandchildren and great-grandchildren, she saw something like a coliseum of thriving, colorful, distinctive life, spectacular precisely for its improbability. If we address the environmental crisis now, the future life we will have enabled—reclaimed, re-freed—to thrive might look the same.

•

It wasn't until 2003 that the question of what Lowell had been seeing and documenting for all those years was finally answered. A retired optometrist, Sherman Schultz, noted that the modifications Lowell had made to his telescope turned it into something quite similar to the tool used to detect cataracts. The tiny aperture, which Lowell felt offered a clearer image of the planets he was observing, cast shadows of his own blood vessels and floaters in the vitreous body of his eye onto his retina, making them visible. By accident, Lowell took a tool that was invented to reveal the things that are farthest from a beholder's eyes and altered it to reveal the things that are closest to them. He was born shortly after the Industrial Revolution—the period during which Western humanity most dramatically imposed its own vision on Earth, altering it forever. The maps Lowell drew of a dying civilization's planet were maps of the structures and imperfections of his own eyes.

The house where I grew up has not shrunk, and neither have my grandmother's hands. Just like Lowell, I misattribute phenomena I observe to external changes rather than internal ones. Even those of us who accept the fact of anthropogenic climate change deny our personal contribution to it. We believe that the environmental crisis is caused by large outside forces and therefore can be solved only by large outside forces. But recognizing that we are responsible for the problem is the beginning of taking responsibility for the solution.

The planet will get revenge on us, or we will be the planet's revenge.

Home Is Almost Always Imperceptible

I am in Brooklyn, sitting on the floor of my son's bedroom as I type these words. He spends virtually no waking time in here, and so, other than when putting away laundry, neither do I. Which is why I can still detect the subtle ways it smells different from the rest of the house: the almost imperceptible mold of the Landmark Books series he inherited from his uncle, the soap and shampoo that are unique to his bathroom, the stuffed-animal scents of bears, pigs, and tigers received for birthdays, won at carnivals, or exchanged for teeth.

Have you ever become suddenly aware of your home's smell? Perhaps upon returning from a long trip? Or because a visitor mentions it? Under normal circumstances, we are literally unable to smell where we live—to smell anything that we are used to. According to the cognitive psychologist Pamela Dalton, after only about two inhales, "the receptors in your nose sort of switch off." Having decided that an odor isn't threatening, we stop paying attention to it. Get an air freshener and see if after a week you find yourself questioning whether it's working. This rapid adaptation to smell is likely an evolutionary one: rather than expend our attention on something that we know is safe, we can direct our resources to detecting new, potentially dangerous stimuli in our environment. Many evolutionary biologists believe this arose from the need to detect when meat was no longer safe to eat.

It seems untrue that this phenomenon applies to sight and sound as well—that we stop hearing something after a few seconds of listening, or stop seeing after a few seconds of looking—but this is exactly what happens. While not as dramatic as with smell, sensory adaptation applies to all the senses. People who live beside construction sites tend not to hear the racket. When you rest your hand on a dog, you at first feel the warmth and fur, but after only a few moments you become unaware of touching anything at all. The sky is in my field of vision for most of the day, but other than when I deliberately focus my attention on something—a daytime moon, a rainbow—I am capable of forgetting that the sky is there at all. What is always there stops being there.

For most people, home is the most familiar, least threatening place. Because of that, it is also the place we are least capable of accurately perceiving.

Glimpses of Home

You have to achieve at least twenty thousand miles of distance from Earth in order to see it as a globe. "The Blue Marble" was not the first photograph of Earth, but it was the first of the fully illuminated whole. The photo that came to be one of the most reprinted and recognizable images not just *of* Earth but *on* Earth was taken on a somewhat illicit impulse. "Photo sessions were scheduled events in a rigorous flight plan that detailed every step essential to success," wrote the filmmaker Al Reinert. "The film itself was strictly rationed like everything else on those perilous flights; there were 23 magazines onboard for the 70mm Hasselblad cameras, twelve color and eleven black-and-white, all intended for serious documentation purposes. They weren't supposed to be looking out the window, either."

Apollo 17 was the last manned mission to the moon, and when the crew reached their destination, they collected the largest number of lunar samples to date. But the images of Earth have proved to be its most enduring contribution to humankind. As the Apollo 8 astronaut William Anders—the man who took "Earthrise," a photograph that preceded "The Blue Marble"—put it, "We came all this way to explore the Moon, and the most important thing is that we discovered the Earth."

Many have attributed the rise of the environmental movement to those first photographs of Earth. Some credit the planet's apparent fragility in the images—alone, unsupported, and suspended in black—for inspiring a collective desire to protect it.

Astronauts have been profoundly moved, and changed, by the sight of Earth from space. It wasn't when he landed on the moon that Alan Shepard cried, but when he looked back at his home planet. The experience is so powerful and consistent among space travelers that it has been given a name, the "overview effect."

Awe is inspired by two things: beauty and vastness. It's hard to imagine anything more transformatively beautiful and vast than the planet as seen from space, especially as it is framed by a seemingly infinite black emptiness. It is perhaps the clearest visual illustration of interconnection, the evolution of life, deep time, and infinity. From this vantage, the "environment" is no longer an environment, a concept, a context, over there, outside of us. It is everything, including us.

The overview effect changes people. One Apollo astronaut became a preacher upon returning to Earth. One began transcendental meditation and devoted himself to volunteering. One, Edgar Mitchell, founded the Institute of Noetic Sciences, which researches human consciousness. "On the return trip home," Mitchell said, "gazing through 240,000 miles of space toward the stars and the planet from which I had come, I suddenly experienced the universe as intelligent, loving, harmonious."

Since Yuri Gagarin became the first man in space, in 1961, only 567 people have seen our home with their naked eyes. Most astronauts have seen Earth only in partial shadow, and the rarity of witnessing the fully illuminated planet is likely what motivated the Apollo 17 crew member to photograph it. According to the space engineer Isaac DeSouza, "540 [now 567] people experiencing space is a novelty. One million people experiencing it is a movement. One billion people, and we've revolutionized how the planet thinks of the Earth." For that reason, he cofounded SpaceVR, a start-up intending to send a satellite equipped with high-resolution virtual-reality cameras into orbit. The company's goal: "to give everyone in the world the opportunity to experience the 'Overview Effect.'"

Commenting on this possibility, the University of Pennsylvania researcher Johannes Eichstaedt noted, "Behavior is extremely hard to change, so to stumble across something that has such a profound and reproducible effect, that should make psychologists sit up straight and say, 'What's going on here? How can we have more of this? . . . In the end, what we care about is how to induce these experiences. They help people in some ways be more adaptive, feel more connected, reframe troubles."

Describing his nonvirtual experience of the overview effect, the astronaut Ron Garan said, "I was flooded with both emotion and awareness. But as I looked down at the Earth—this stunning, fragile oasis, this island that has been given to us, and that has protected all life from the harshness of space—a sadness came over me, and I was hit in the gut with an undeniable, sobering contradiction."

What contradiction? That our planet protects us from the harshness of space but we don't protect it from the harshness of us? That while everyone knows we live on Earth, you can believe it only by leaving?

Glimpses of Ourselves

The earliest spectacles, made in Pisa, date back to around 1290. A decade later, in Venice, the convex glass mirror was invented—likely an accidental discovery, connected with the development of the lenses used in glasses. The rare mirrors that existed before were dull, imprecise, and distorted. Just as a journey to the moon enabled us to see our own planet, an invention meant to help us see others enabled us to see ourselves.

While the first clear images of Earth inspired its inhabitants to protect it, kick-starting the environmental movement, the first clear reflections of our ancestors inspired them to understand themselves. By 1500, a wealthy person could afford a mirror. "As the fourteenth century drew to a close and people started to see themselves as *individual* members of their communities," writes the historian Ian Mortimer, "they started to emphasize their *personal* relationships with God. You can see that transformation reflected in religious patronage. If in 1340 a wealthy man built a chantry chapel to sing Masses for his soul, he would have the interior decorated with religious paintings, such as the adoration of the Magi. By 1400, if the founder's descendant redecorated that chapel, he would have himself painted as one of the Magi." The rise of the glass mirror also precipitated a rise in self-portraiture (which might be considered early selfies) and first-person novels, and intensified personal reflection in letters.

When babies begin to recognize their reflections, they display avoidance, withdrawal, and embarrassment, perhaps best exemplifying the term "self-conscious."

Only a few nonhuman species are known to recognize their reflections in mirrors. These include orcas, dolphins, great apes, elephants, and magpies. A recent addition to this list is a species of tiny coral fish called the cleaner wrasse, so-named because it eats mucus, parasites, and dead skin off larger fish. Usually, scientists test mirror recognition by placing a dot on an animal's face and seeing if the animal engages with it, making the connection between its face and its reflection. To test the cleaner wrasse, scientists placed individuals in tanks with mirrors. At first, the fish behaved aggressively, attacking their reflections. "But eventually," reports *National Geographic*, "this behavior gave way to something far more interesting." The fish started "approaching their reflections upside-down, or dashing towards the mirror quickly, only to stop right before touching it. At this phase, the researchers say, the cleaner wrasse were 'contingency-testing'—directly interacting with their reflections, and perhaps just starting to understand that they were looking at themselves and not another wrasse." After the fish got used to the mirrors, scientists injected some of them with a colored gel that could be seen under their skin—a change they could detect only if they looked at their reflections. Some were injected with a gel that did nothing to their skin, and others were injected with the colored gel but were not offered mirrors. "Fish injected with a clear mark didn't scrape, and neither did those with a colored mark when no mirror was present. Only when the fish could see their mark in a mirror did they try to scrape it off, suggesting that they recognized their reflections as their own bodies."

Cleaner wrasse live in the kinds of reefs that will become extinct even if we successfully meet the Paris accord's goals and warm the earth by only two degrees.

•

About a decade after "The Blue Marble" circulated around the globe, incontrovertible evidence of anthropogenic global warming emerged. In 1988, the NASA scientist James Hansen testified before the U.S. Senate Committee on Energy and Natural Resources. "Global warming," he said, "has reached a level such that we can ascribe with a high degree of confidence a cause and effect relationship between the greenhouse effect and the observed warming." His testimony helped to upload the term "global warming" into the American vernacular. That same year, then-presidential nominee George H. W. Bush, a self-identified environmentalist, gave a speech in Michigan, the car capital of America, in which he said, "Our land, water and soil support a remarkable range of human activities, but they can only take so much and we must remember to treat them not as a given but as a gift. These issues know no ideology, no political boundaries. It's not a liberal or conservative thing we're talking about." He pledged to "fight the greenhouse effect with the White House effect." That year, forty-two senators—about half of whom were Republicans—urged Reagan to press for an international treaty modeled on the ozone agreement.

It's worth revisiting the ozone agreement, if only because it demonstrates the possibility of international environmental cooperation. Signed in 1987, it was called the Montreal Protocol, and its original version required developed countries to begin phasing out chlorofluorocarbons—ozone-destroying compounds found in refrigerants and aerosol propellants—in 1993 and achieve a 50 percent reduction by 1998. They were also required to halt their production and consumption of halons, compounds used in fire extinguishers that damage the ozone layer. According to the EPA, "Because of measures taken under the Montreal Protocol, emissions of ODS {ozone-depleting substances} are falling and the ozone layer is expected to be fully healed near the middle of the 21st century."

About six years before James Hansen's congressional testimony, and after a decade of investment in climate change research, Exxon slashed its budget for investigating how CO_2 emissions from fossil fuels would affect the planet, reducing it by 83 percent. Then the fossil fuel industry launched a disinformation campaign, producing false reports that, if believed, would excuse America from a painful self-examination. In his investigative article "Losing Earth: The Decade We Almost Stopped Climate Change," Nathaniel Rich writes: "It is incontrovertibly true that senior employees at the company that would later become Exxon, like those at most other major oil-and-gas corporations, knew about the dangers of climate change as early as the 1950s. But the automobile industry knew, too, and began conducting its own research by the early 1980s, as did the major trade groups representing the electrical grid. They all own responsibility for our current paralysis and have made it more painful than necessary. But they haven't done it alone. The United States government knew . . . Everybody knew."

And yet we still displayed avoidance, withdrawal, and embarrassment. We were—and to some extent remain—in the earliest stages of development when it comes to examining our impact on our planet: babies recognizing ourselves in the mirror.

In the first one hundred days of George W. Bush's presidency—thirteen years after his father's speech in Michigan—he retracted a campaign promise to regulate emissions from coal-fired power plants and withdrew the United States from the Kyoto global climate change treaty. His justification was as significant as the withdrawal itself: he cited scientific doubt. Bush promised that "[his] Administration's climate change policy will be science-based." That same year, he established the U.S. Climate Change Research Initiative, one of whose chief priorities was to study "areas of uncertainty" in climate change science. In his speech discussing why the United States would not take part in the Kyoto Protocol, Bush said, "We do not know how much effect natural fluctuations in climate may have had on warming. We do not know how much our climate could or will change in the future. We do not know how fast change will occur, or even how some of our actions could impact it."

In America, it is easier than ever for the Left to blame the Right for our environmental negligence, especially now that we have a president who shrinks national forests, opens protected land to oil interests, renders the Environmental Protection Agency a Fossil Fuel Protection Agency, tries to defibrillate the coal industry, removes federal protection of waterways, and pulls out of the Paris accord. But that blaming can also be a means of turning away from our own reflections. Although his administration achieved some environmental progress, Obama failed to push climate legislation forward during his first two years in office, when he had a Democratic Congress. Recently, supposedly progressive hotbeds have failed on climate change: Washington State rejected a carbon tax, and Colorado refused to slow down oil and gas projects. Abroad, the French turned out in huge numbers to protest a gasoline tax. After three weeks of violent demonstrations, Emmanuel Macron announced that the tax would be suspended.

Signs of progress like We Are Still In (a coalition of American leaders committed to achieving the goals of the Paris accord without the help of the federal government), the Last Plastic Straw, Meatless Mondays, the plastic-bag tax, and even China's 2020 action plan for pollution and climate change—are they all just contingency-testing? Are we merely experimenting with how our behavior affects our reflections, as the cleaner wrasse did before making the connection? Just starting to understand that we are looking at ourselves and not governments or corporations? These are first steps, certainly, but they are only baby steps. And we need to be sprinting toward change.

Almost fifty years after the Apollo 17 astronauts took "The Blue Marble," and about thirty years after James Hansen first testified on global warming, America elected a president who has posted more than one hundred tweets expressing climate change skepticism, including these:

> We should be focused on magnificently clean and healthy air and not distracted by the expensive hoax that is global warming!

> They call it "climate change" now because the words "global warming" didn't work anymore. Same people fighting hard to keep it all going!

> This very expensive GLOBAL WARMING bullshit has got to stop. Our planet is freezing, record low temps, and our GW scientists are stuck in ice.

What is your response to those statements? Anger? Terror? Defiance? They fill me with a primitive rage that I feel only when someone endangers my children.

But those responses are misplaced.

There is a far more pernicious form of science denial than Trump's: the form that parades as acceptance. Those of us who know what is happening but do far too little about it are more deserving of the anger. We should be terrified of ourselves. We are the ones we have to defy. Self-recognition does not always indicate self-awareness, critics of the mirror test say. I am the person endangering my children.

Mortgaging the Home

"I am convinced that humans need to leave Earth," Stephen Hawking said. "The Earth is becoming too small for us, our physical resources are being drained at an alarming rate."

The Global Footprint Network (GFN) is a consortium of scientists, academics, NGOs, universities, and tech institutions that measures the human ecological footprint. By looking at the natural resources required to produce what we consume, as well as how much greenhouse gas is emitted, the GFN calculates a budget that lets us know to what extent we are living within our means. The answer depends entirely on who is referred to by "we." If the 7.5 billion people on the planet had the needs and outputs of the average Bangladeshi, we would require an Earth the size of Asia to live sustainably—our planet would be far more than enough for us. Earth is approximately the right size to supply the Chinese budget—despite being the face of environmental villainy, the Chinese currently get the balance right. For everyone to live like an American, we would need at least four Earths.

According to the GFN, the end of the 1980s marked the

end of Earth's ability to supply earthlings' demands. From that point forward, we have been living in what might be called ecological debt—spending at an unsustainable rate. The GFN estimates that by the 2030s, we will have reached the point of requiring a second Earth to satisfy our earthly needs.

Most readers of this book live with some kind of debt, whether it be student loans, auto loans, credit card debt, or a home mortgage. (Seventy-three percent of American consumers have outstanding debt when they die.) When considering a loan, banks look at the applicant's debt-to-income (DTI) ratio. Most financial planners consider a debt-to-income ratio of 36 percent or lower to be healthy. No one with a DTI ratio of higher than 45 percent is likely to get a loan from a bank. (A key part of the Dodd-Frank Act, a response to the financial crisis of 2008, was the qualified-mortgage rule, which stipulated that borrowers must have a DTI ratio of 43 percent or less to qualify for a loan.) Humanity has a DTI ratio of 150 percent, meaning we are consuming natural resources at a rate 50 percent greater than Earth's ability to replenish them.

The expression "mortgaging our children's future" has been used in many contexts, from tax cuts that will produce debt to a lack of investment in infrastructure. Someone will have to pay for our choices, we know without believing. We are also mortgaging our children's future with lifestyles that will create future environmental calamities. In fact, twenty-one youth plaintiffs have filed a "constitutional climate lawsuit" against the federal government, asserting that "through the government's affirmative actions that cause climate change, it has violated the youngest generation's constitutional rights to life, liberty, and property, as well as failed to protect essential public trust resources." The Trump administration tried to intervene and get the case dismissed, but the Supreme Court ruled unanimously in favor of the youth plaintiffs, allowing the case to proceed.

The American dream is to have a better life than one's parents—better primarily in the sense of affluence. My grandparents lived in a larger and more valuable home than that of their parents. My parents live in a larger and more valuable home than that of their parents. I live in a larger and more valuable home than that of my parents. This defining of "having enough" as "having more" is the mentality that created both America and global warming. It is problematic on all scales, and self-destruction is built into the model, because nothing can grow forever. Many economists argue that millennials are the first generation of Americans since the Great Depression to do worse, financially, than their parents.

My grandmother and I used to arrange coins into paper rolls to take to the bank and exchange for bills; if we left with five dollars, we were rich. On supermarket trips, she bought sale-priced foods as if shopping not only for her living family but for all her dead relatives. When taking me out for breakfast—a treat reserved for special occasions—she would buy two bagels, one with cream cheese, and then transfer half the cream cheese onto the dry bagel. And when she retired, after decades of twelve-hour days managing corner grocery stores, she had more than half a million dollars in savings. She wasn't putting away all that money so she could leave it to her children and grandchildren. She simply wanted to make sure that she would never have to take anything from us—that no one would ever have to pay for her care.

My great-grandparents lived in a wooden house with no indoor plumbing and on cold nights would sleep on the kitchen floor by the stove. They never could have believed the things I have: a car that I drive for convenience rather than necessity, a pantry stocked with foods imported from all over the planet, a home with rooms that aren't even used on a daily basis. And my great-grandchildren won't believe it, either, although their disbelief will have a different spirit: How could you have lived so high and left us with a bill too large to be paid—too large to be survived?

Debts caused by tax cuts can be negotiated. Infrastructure that has fallen into disrepair can be fixed or replaced. Even many forms of environmental damage—ocean dead zones, water pollution, biodiversity loss, deforestation—can be and have been reversed. But with respect to greenhouse gas emissions, the notion of mortgaging doesn't make sense: no one—no institution, no god—would give us a loan so wildly out of proportion with our means. And while humankind might feel too big to fail, no one will bail us out.

A Second Home

It's entirely possible that we will have our necessary second Earth one day. People like Stephen Hawking have argued that we need to start colonizing space within one hundred years to keep the species alive, and people like Elon Musk are actively working toward making it a reality. We may figure out how to launch one hundred thousand people at a time (the alignment of the planets allows for favorable departures only once every two years—according to Musk, the Mars colonial fleet would launch en masse, "kind of like *Battlestar Galactica*"), devise a way to manufacture rocket fuel on Mars, and solve the problem of building infrastructure necessary to sustain a colony, not to mention make a home in a place with temperatures of minus eighty degrees Fahrenheit (talk about a climate problem) and deadly radiation. If we can't clean up our water and air, we can always manage on a planet without any.

A mere sixty-six years separate the Wright brothers' first flight and Neil Armstrong's first step on the moon—a span shorter than it took Noah to build the ark, and shorter than my parents' lives. If someone in the time of the Wright brothers had suggested that in fewer than seven decades there would be a human on the moon—forget about the hundreds of millions of earthlings watching it on television sets in homes—the notion would likely have been met with something stronger than skepticism. Humanity has a tendency to underestimate its own power to create and destroy.

But maybe the question isn't *can* we do it (let's assume we can), or even *should* we (let's assume it could be accomplished, as Musk has predicted, with relatively little investment and relatively little time), but rather what, beyond watching and hoping, should we do in the meantime. To what extent does this deus ex machina—or the dozens of other engineering strategies that have been proposed, from blocking sunlight with the mass injection of sulfate aerosols, to post-facto carbon removal, to the "engineered weathering" of the oceans—deserve our attention?

And what portion of that attention should conclude in fear of Frankenstein solutions that turn against their creators? Responding to techno-interventions in the environment, the usually restrained National Academy of Sciences declared, "There is significant potential for unanticipated, unmanageable, and regrettable consequences in multiple human dimensions from [attempts to modify the climate], including political, social, legal, economic, and ethical dimensions." How many of our eggs should we put in the basket of miracle fixes?

And how many in the basket of legislated change? Can't we—*they*—just tax fossil fuels in proportion to the amount of carbon released? Institute an ambitious cap-and-trade program? Incentivize international cooperation via tariffs? Regulate global emissions in a way that not even a recalcitrant leader can exempt a country from?

We can try. We have to try. When it comes to working

against the destruction of our home, the answer is never *either/or*—it's always *both/and.* We can no longer afford to pick and choose which planetary illnesses we attempt to remedy, or which remedies we attempt. We must strive to end the extraction and burning of fossil fuels, and invest in renewable energy, and recycle, and employ renewable materials, and phase out hydrofluorocarbons in refrigerants, and plant trees, and protect trees, and fly less, and drive less, and advocate for a carbon tax, and change our farming practices, and reduce food waste, *and* reduce our consumption of animal products. And so much more.

But technological and economic solutions are good at fixing technological and economic problems. While the planetary crisis will require invention and legislation, it is a far broader kind of problem—an *environmental* problem—that involves social challenges like overpopulation, the disempowerment of women, income inequality, and consumption habits. It reaches into not only our future but our past.

According to Project Drawdown, four of the most effective strategies for mitigating global warming are reducing food waste, educating girls, providing family planning and reproductive healthcare, and collectively shifting to a plant-rich diet. The benefits of these advancements extend far beyond the reduction of greenhouse gas emissions, and their primary cost is our collective effort. But there is no getting around that cost.

Civilian efforts during World War II were indispensable for defeating the enemies abroad, but they also triggered social progress at home. Despite the injustice that many American minorities suffered during the war—segregated armies, the abuse of Japanese Americans—it was also a period of social progress that would shape American culture. In 1941, Roosevelt signed Executive Order 8802, which outlawed racial discrimination in national defense industries and in government. Membership in the NAACP increased from eighteen thousand to almost five hundred thousand during the war. In the South, the percentage of African Americans who registered to vote jumped from 2 to 12 percent, and many referred to the war as a "Double V"—a victory abroad and a victory over segregation at home. The exodus of men for the battlefront cleared a space for nearly seven million women to join the industrial workforce. Jobs opened for Mexican Americans as well; between 1941 and 1944, the number working in Los Angeles shipyards increased from zero to seventeen thousand. These newfound opportunities for women and minorities exposed structural prejudice, cultivated professional skills, and galvanized civil rights movements to come.

Saving ourselves will require collective action, and acting collectively will change us—especially if we change not because we are inspired to, not because we "see the light," but rather because, sensing an approaching dark, we compel ourselves to act on knowledge that we can't believe. When a couple suffers a betrayal—an affair, for example—and is deciding whether to stay together, the famed therapist Esther Perel encourages the partners to think of their marriage in these terms: "Your first marriage is over. Would you like to create a second one together?"

Perhaps we don't need to abandon home to save ourselves: our second Earth could be a transformed version of the one we currently inhabit. Either way, we are going to have to live on a new planet—one that we reach by leaving, or one that we reach by staying. Those two forms of saving ourselves would say very different things about us.

What kind of future would you predict for a civilization that abandons its home? We would be revealed by that decision, and we would be changed by it. People who think of home as dispensable will be able to think of anything as dispensable, and will become a dispensable people.

What kind of future would you predict for a civilization that acts collectively to save its home? We would be revealed by that decision, and we would be changed by it. By making the necessary leap—which is not a leap of faith but of action—we would do more than save our planet. We would make ourselves worthy of salvation.

Glass

The Hubble Space Telescope was launched in 1990. Its optics are so powerful and precise that if it were aimed at Earth—and able to overcome the haziness of the atmosphere and blurring speed of the telescope's orbit—it could read this page over your shoulder. Turned away from Earth, Hubble can see nearly to the beginning of time.

Hubble was originally funded in the 1970s but took twenty years to design, build, and launch. The modern equivalent of a cathedral, it is a physical expression of humankind's collective achievements and ambitions. Among its many accomplishments is determining the size and age of the universe, detecting the first organic molecule outside our solar system, revealing that nearly all galaxies contain supermassive black holes, understanding how planets are born, and witnessing a distant supernova that suggests the universe only recently began speeding up.

It was almost all for naught. When the first images came through, it was obvious that there was a serious problem with the optics. The mirror—arguably the most precisely carved mirror ever made—was ground too shallow by about one-fiftieth the width of a human hair. Instead of bringing 70 percent of a star's light into the focal point, Hubble could manage only about 10 percent. The images were a disappointment, and the project became NASA's worst-ever embarrassment—in *Naked Gun 2½*, a photo of Hubble is hung alongside those of the Hindenburg and Michael Dukakis.

This was not an easy problem to solve. The mirror couldn't be re-polished in space. Neither could a replacement mirror be installed in orbit. And it would be far too expensive to bring Hubble back to Earth for repairs. The saving grace, as it turned out, was the precision of the error—an optical component with the same degree of error in the other direction could correct the focus. In 1300, the effort to create glasses led to the invention of the mirror; seven centuries later, the most sophisticated mirror ever made needed a pair of glasses. Sometimes, even the most vast and complex problems can be solved with a simple correction, a balancing. We don't need to reinvent food but to uninvent it. The future of farming and eating needs to resemble the past.

•

Vincenzo Peruggia had been hired by the Louvre to construct protective glass cases for a number of paintings. On the evening of August 20, 1911, he and two accomplices hid inside a closet used for storing student art supplies. When they emerged the next morning, Peruggia went straight to the *Mona Lisa*, removed it from the wall, and carried it out the museum's main entrance.

At the time, the *Mona Lisa* wasn't widely known outside the art world; it was not the most famous work in its gallery, much less the museum. It took twenty-four hours before the painting's absence was even noticed. But once it drew the attention of the burgeoning print media, the theft became international intrigue, and the *Mona Lisa*, now referred to as a masterpiece, became the most famous painting in the world. When the Louvre reopened, after a week of being closed for investigation, queues formed outside for the first time in the museum's history. In the two years between the painting's theft and return, more people came to see the bare wall on which it had hung—"the mark of shame"—than had ever come to see the painting.

Franz Kafka paid the empty wall a visit within a month of the painting's disappearance, the absence now among his collection of "invisible curiosities"—sights, events, people, and works of art that he had missed seeing. The following year, perhaps inspired by the experience, Kafka wrote his masterpiece, *The Metamorphosis*, in which a man awakes one morning as an insect, his perspective radically altered, and his body—his first home—no longer hospitable.

The fame of the painting has only increased with time—or perhaps it's more correct to say that the fame of the fame of the painting has increased. People want to see the *Mona Lisa* because other people want to see the *Mona Lisa*. The Louvre estimates that 80 percent of those who visit the museum come to see only that one work. It now resides behind 1.52-inch-thick bulletproof glass. While the purpose of the glass is to protect the world's most valuable painting, the effect is to enhance our sense of its value and vulnerability. When we look at the *Mona Lisa*, the bulletproof glass also serves as a corrective lens.

•

I owned a pair of glasses for two full years before I wore them regularly. On one of my kindergarten field trips, the teacher told all the kids to move to one side of the bus. There was something to see out the windows. The bus leaned under the shifted weight, and the other kids gasped dramatically.

"Do you see it?" my teacher asked over my shoulder.

"See what?"

"If you were wearing your glasses, you would see it."

"I don't know what I'm supposed to be seeing."

"If you wore your glasses, you would."

At the time, I suspected that everyone was in on a trick, that the students went to the side when signaled, pointed and gasped at nothing—all to teach me a lesson.

The next day, my teacher said I looked handsome in the aviator glasses my mother had picked out for me, but I knew the truth. I asked her what they'd been looking at on the bus.

"A daytime moon," she said.

"But I was looking at a window washer," I said. "He was really small."

"We were looking at the moon."

"Of *course* I could have seen that."

"But you didn't."

I couldn't see it because I wasn't looking for it. We can wear glasses to correct our vision *on* Earth, and we can go to space to correct our vision *of* Earth. But no glasses or interstellar journey can aim us in the right direction. We direct our gazes at the things we want to see, the things we care about. Our perception is sharpest when our care is heightened, and living things care most when they are afraid. People stared at the *Mona Lisa* after it was stolen. I stared at the window washer because I had a fear of heights.

My hearing is sharpest when listening for a sleeping child. My palate is most refined when I have been asked to determine if a food has gone bad. My vision is sharpest when

I'm in fight-or-flight mode. People often remember near-death experiences as having happened in slow motion, with all their senses heightened. Perhaps this is just another version of hysterical strength.

The problem is, our relationship to the planet is a near-death experience that doesn't feel like one. If we could believe that our planet was in danger, we could see it for what it is. It might be true that if a billion people could experience the overview effect, it would revolutionize how earthlings think of and treat Earth. But the only scenario in which that is likely to happen is when we are on our way to a new home. Imagine that: the entire species shifting to see out the windows of one side of the spacecraft, looking through a thick pane of protective glass, and realizing that our home was a masterpiece.

Jumping from the Golden Gate Bridge results in death 98 percent of the time. More than sixteen thousand have jumped. Among the few survivors, all who have shared their experiences describe changing their minds as soon as they let go. Perhaps our species would experience something similar. Kevin Hines was eighteen when he leaped. If we were to lose our planet, perhaps each of us would think, as Hines did, watching the bridge recede as he fell, "What have I done?"

First Home

"The human race has existed as a separate species for about 2 million years. Civilization came about 10,000 years ago, and the rate of development has been steadily increasing. If humanity is to continue for another million years, it relies on boldly going where no one has gone before . . . We will need to take the practical means of establishing a whole new ecosystem that will survive in an environment that we know very little about and we will need to consider transporting several thousands of people, animals, plants, fungi, bacteria and insects."

So said Stephen Hawking.

If your house were in need of repairs, even extensive ones, would it be bold to abandon it for a new house? What if the new house were sure to be vastly less hospitable and far from everything you've known?

Instead of traveling beyond the horizon, we could venture into our own consciences and colonize still-uninhabited parts of our internal landscapes. Instead of leaping to the distant fantasy of transporting animals on spaceships to other planets, we could start, right now, raising far fewer of them on the extraordinary planet we already have.

When Americans turned off their lights during WWII, they weren't protecting their houses—the blackouts had little utilitarian value—they were protecting their *homes*. They were demonstrating solidarity, and therefore protecting their families and cultures, their safety and freedom.

In a 2016 public service announcement for a Swedish nonprofit, Hawking said, "At the moment, humanity faces a major challenge, and millions of lives are in danger." He then proceeded to speak about obesity and why humankind needs to eat less and get more physical activity. "It's not rocket science."

Millions of lives might be in danger because of eating too much food, but every human life is in danger because of eating too many animal products. It's not rocket science in the colloquial sense, and the answer is not literally rocket science. If we don't demonstrate solidarity through small collective sacrifices, we will not win the war, and if we do not win the war, we will lose the childhood home of every human who has ever lived.

Final Home

I am sitting at my grandmother's bedside. My older brother urged me to come down this weekend. I knew what he meant. For the same reason that my brother didn't say, "She's about to die," I have had a hard time saying what I mean to my grandmother. Even touching her is difficult. I am capable of "I love you," but not "I am going to miss you." I am capable of kissing her hello and goodbye, but not taking her hand while with her.

Looking at my grandmother from this distance, I feel something like the overview effect: home is suddenly vulnerable, beautiful, singular. And I suddenly see all of her at once—in the context of my life, my family, history. Framed by a seemingly infinite black emptiness, my grandmother is in need of, and deserving of, protection.

I punish myself by remembering all the times I took her for granted, or worse. I made faces to my brother while on compulsory phone calls with her. I begged not to have to sleep over at her house, and while there, I watched hours of reruns and spoke to her hardly at all. I turned my face away from her kisses.

Having my own children now, I know that I was doing what

children do. It is not a child's responsibility to take care of an elder (or a home, or a planet); it is an adult's. And that is what my parents have done, bringing her here, making their home her home. They installed a stair lift so she could move between floors, hired occasional and then full-time help, and have never once mentioned how much less privacy they have, or how many more emotional, logistical, and financial responsibilities. Their care for her—which has required many kinds of sacrifice—has been a revelation for me. I don't believe it is a coincidence that this book began to germinate when she moved in with them.

My older son is about to have his bar mitzvah, the Jewish rite of passage into adulthood. Among other things, it marks the transition from being a recipient of what the world has to offer to being a participant in the maintenance and creation of what the world offers others. It is both gorgeous and devastating. He can prepare dinner. He can read himself to sleep. It is complicated, and often painful, to take care of something that you are in the process of letting go of. My son needs me as much as my grandmother needs my parents. But it is also my job, and the job of my parents, not to hold on.

Whether or not we address climate change, we will need to learn to let go. Even if we were to reduce carbon emissions to zero tomorrow, we would continue to witness and experience the effects of our past actions. The planet will not be a home for our children and grandchildren as it was for us—not as comfortable, beautiful, or pleasurable. As Roy Scranton argues in his *New York Times* essay "Learning How to Die in the Anthropocene," it is important to come to terms with that loss:

> The biggest problem climate change poses isn't how the Department of Defense should plan for resource wars,

> or how we should put up sea walls to protect Alphabet City, or when we should evacuate Hoboken. It won't be addressed by buying a Prius, signing a treaty, or turning off the air-conditioning. The biggest problem we face is a philosophical one: understanding that this civilization is *already dead.* The sooner we confront this problem, and the sooner we realize there's nothing we can do to save ourselves, the sooner we can get down to the hard work of adapting, with mortal humility, to our new reality.

Looking at my grandmother, I really understand what he means. There is a sense in which she is *already dead*—hard as that is to write—and accepting her absence is not only the most honest approach but the one that will allow us to fully value her presence.

It is custom, on Yom Kippur, the Jewish Day of Atonement, to say the Mi Shebeirach prayer on behalf of an ill loved one:

> *May the One Who Blesses*
> *overflow with compassion upon her,*
> *to restore her,*
> *to heal her,*
> *to strengthen her,*
> *to enliven her.*
> *The One will send her, speedily,*
> *a complete healing—*
> *healing of the soul and healing of the body—*
> *along with all the ill,*
> *among the people of Israel and all humankind,*
> *soon,*
> *speedily,*

without delay,
and let us all say: Amen!

After days of deliberation, my mother chose not to say the prayer on my grandmother's behalf this year. My grandmother is not going to heal, and *shouldn't* heal. She is ninety-nine years old. She is in no pain, physical or emotional. It would be cruel to extend the duration of her life at the expense of her experience of her life.

It is true that there is nothing we can do to "save" my grandmother. It is also true that we can save things that matter—to her and to us. She can spend her remaining time in a peaceful setting. My parents bought her a special mattress that helps prevent bedsores. They moved her to the window so she could see the tree and feel the sunlight. They hired a live-in nurse, for medical care but also so she'll never be alone. They spend hours every day talking to her and encourage her grandchildren to come as often as possible, and her great-grandchildren to FaceTime. They give her the things that make her happy: chocolate, photographs of her family, recordings of the Yiddish songs she listened to as a child, company.

We cannot save the coral reefs. We cannot save the Amazon. It's unlikely that we'll be able to save coastal cities. The scale of inevitable loss is almost enough to make any further struggle feel futile. But only almost. Millions of people—perhaps tens or hundreds of millions—will die because of climate change, and the number matters. Hundreds of millions of people, perhaps billions, will become climate refugees. The number of refugees matters. It matters how many days per year children will be able to play outside, how much food and water there will be, how many years average life expectancies will shed. These numbers matter, because they are not just numbers—each corresponds to an individual, with a family, and idiosyncrasies, and phobias, and allergies, and favorite foods, and recurring dreams, and a song stuck in her head, and a singular handprint, and a particular laugh. An individual who inhales molecules that we have exhaled. It is hard to care about the lives of millions; it is impossible not to care about one life. But maybe we don't need to care about any of them. We just need to save them.

I do not believe that the biggest challenge that climate change poses is a philosophical one. And I'm quite sure someone in sub-Saharan Africa, or South Asia, or Latin America—where climate change is already painfully felt—would agree with me. The biggest challenge is to save as much as we can: as many trees, as many icebergs, as many degrees, as many species, as many lives—*soon, speedily, and without delay.*

That we want everybody on Earth not only to have a healthy life but to feel at home should go without saying. But it doesn't. It requires not only saying but repeating. We must force ourselves to face the mirror, and force ourselves to look. We must engage in perpetual disputes with ourselves to do what needs to be done. "Listen to me," implores the soul in the first suicide note, when it begins to make its case for life. "Behold, it is good for men to listen."

IV. DISPUTE WITH THE SOUL

I don't know.

What's not to know?

I don't know how I've gotten this far—learned this much, convinced myself this thoroughly of the need to change—and yet still doubt that I'll change. Are you hopeful?

That you'll change?

That humankind will figure this out.

We've already figured it out.

That we will act on what we've figured out.

Have you noticed how often conversations about climate change end with the question of hopefulness?

Have you noticed how often conversations about climate change end?

That's because we feel hopeful and are comfortable putting off the discussion.

No. It's because we feel hopeless and are uncomfortable discussing it.

Either way, it's hope that allows the subject of climate change to be eclipsed—in news and politics, in our lives—by more "urgent" issues. If you were a doctor, would you ask a cancer patient if he was hopeful?

I might. Positivity seems to improve recovery.

If you were a doctor, would you ask a cancer patient if he was hopeful without also asking what course of treatment he planned to take?

No, probably not.

And what if he told you he planned to do nothing at all? Would you ask him if he was hopeful, then?

I might ask if he was depending on a miracle or just accepting death.

Right. If someone faces a life-threatening crisis, and chooses not to address it, asking if he is hopeful is shorthand for asking if he is depending on a miracle or just accepting death.

There have been moments when writing this book has made me hopeful, but almost always I've felt rage or despair.

You've been stealing those pleasures.

Hope?

Yes, but also rage and despair.

Stealing?

Not giving anything in return.

Rage and despair are pleasures?

The guiltiest. Why do you think New York *magazine's doomsday article about global warming went viral? People were suddenly ravenous for climate science?* No, *we were ravenous for a vivid description of our apocalypse. We're drawn to it the same way we're drawn to horror movies, car accidents, and the chaos of the current administration. And don't pretend that the bleakest scenarios aren't your favorite parts to write.*

I'm not pretending.

Admit it: it feels good to point out other people's shortcomings.

That's not fair.

Not at all. So pay for all the pleasures you've stolen.

I've spent two years writing this book, trying to persuade as many people as I can to change their lives. Isn't that something?

Not enough.

What would be enough?

Change your own life.

I know.

But?

I don't know.

What's not to know?

Is there anything more narcissistic than believing the choices you make affect everyone?

Only one thing: believing the choices you make affect no one. As you've spent the last three sections explaining.

Maybe that was all just writing. More stolen pleasure. Climate change is a problem on the scale of China and ExxonMobil. Just one hundred companies are responsible for 71 percent of greenhouse gas emissions. Putting the onus on individuals isn't fair.

If you were a child, your obsession with fairness would be admirable.

Forget about fairness. Making this about individuals is naive in terms of what needs to be done, while letting politicians and businesses off the hook.

But companies produce what we buy; farmers grow what we eat. They commit crimes on our behalf. On top of which, while a lot of people talk about how climate change is a problem of nations and corporations, no one seems to have a plan for affecting policy change among nations and corporations. And decrying the bad guys is no more of an action than marching with the good guys.

"We have to do something." It's the phrase that seems to be on the tip of everyone's tongue these days, the unofficial slogan of our moment. And yet almost no one does anything beyond reiterating the need to do something. We either don't know what to do or don't want to do it. So instead we stagger across the battlefield, firing blank after blank without aiming: something, something, something . . .

But there *is* something we can do. Choosing to eat fewer animal products is probably the most important action an individual can take to reverse global warming—it has a known and significant effect on the environment, and, done collectively, would push the culture and the marketplace with more force than any march.

So there it is.

I don't know.

What's not to know?

I've already been changed by the process of writing this book. I can imagine doing radio and print interviews, writing op-eds, giving readings in cities around the world. I can imagine stealing the pleasure of righteousness and then coming back to my hotel after one of those events and eating a burger behind a locked door—stealing that pleasure, too. Can you think of anything more pathetic?

It's not a great image. But I can think of many scenarios more pathetic—like if you didn't bother with the truth, or were too afraid to learn it. Or if you knew the truth but didn't care, or wouldn't make an effort in response. Or if you tried but felt no remorse when you failed.

It's always driven me crazy that my friend—a fellow writer and, what's more, a passionate environmentalist—has refused to read my book *Eating Animals*. It upsets me because he is a sensitive thinker who cares and writes about the preservation of nature. If *he* is unwilling even to learn about the connection between eating and the environment, what hope is there for hundreds of millions of people to alter their lifelong habits?

Why won't he read it?

He told me he's afraid to read the book because he knows that it will require him to make a change he can't make.

Congratulations, you're better than your friend. Pointing out his shortcomings must have soothed your guilt about your own. And while we're on the subject of your narcissism, why are you making your patheticness the subject here?

I was using his shortcomings to *illustrate* my own: if I argue against eating animal products while continuing to eat them myself, then I am a massive hypocrite.

Why is that important to say?

No one wants to be a hypocrite.

So be perfect instead.

Don't do that.

What?

Be glib about the real pain involved in trying to do the right thing.

Don't do that.

What?

Make your emotions more urgent than the planet's destruction.

Our emotions—and lack of emotions—are destroying the planet.

Without a doubt. You don't want to give up your burgers, your drives to the grocery store and flights to Europe, your cheap electricity. You don't want to make the dinner party awkward, or have anyone think you're a drag or, worse, an asshole. You don't do something *because you just don't feel like it. But as ever, you have your comfort to protect, so you convince yourself that knowing about it—writing a book about it—*is *doing something.*

So you're . . . *not* hopeful?

You're entirely capable of doing things you aren't moved to do and refraining from things that you want to do. That doesn't make you Gandhi. It makes you an adult.

That really isn't fair.

From the mouth of a child. Do you know why ostriches bury their heads in the sand?

Because they think no one can see them if they can't see anyone.

Pretty dumb, right? Except ostriches don't *bury their heads in the sand—they bury their eggs for warmth and protection, and occasionally submerge their heads to turn them. Humans look at ostriches caring for their offspring and mistake it for stupidity. But* we *are the animals who assume that the world goes dark when we close our eyes. Mistaking avoidance for safety happens to be one of the most effective ways to kill our offspring. So is mistaking knowledge for action. No one wants to be a hypocrite, but isn't blinking every now and then better than pressing your eyes shut? The important measurement is not the distance from unattainable perfection, but from unforgivable inaction.*

I don't know.

Let me ask you a question: What's the opposite of someone who leaves lights on in empty rooms, buys inefficient appliances, and pumps the air-conditioning even when no one is home?

Someone who is attentive to energy use?

And what's the opposite of someone who takes cars everywhere he goes, no matter the distance, and no matter the convenience of public transportation?

Someone who is attentive to how much he drives?

What is the opposite of someone who eats a lot of meat, dairy, and eggs?

A vegan.

No. The opposite of someone who eats a lot of animal products is someone who is attentive to how often he eats animal products. The best way to excuse oneself from a challenging idea is to pretend there are only two options.

You wrote about Frankfurter's response to Karski as if he had only two options. Perhaps belief really is all or nothing, but what about action? Couldn't Frankfurter have done something *with what he knew to be true? Perhaps he wasn't going to starve himself in front of the White House, dying a slow death while the world looked on. But surely he could have convened a group of influential figures to hear what Karski had to say, or urged Congress to open an official investigation into German atrocities, or simply used his voice to publicly raise the urgent questions?*

We can imagine his struggles to believe Karski during that meeting, but what about when, only a couple of years later, he saw the first images of the concentration camps? Do you think he believed what he was seeing then? And when he looked into the sunken eyes of those starved fathers and mothers, the heaps of dead sons and daughters? When the Supreme Court justice judged himself, do you think he felt complicit in genocide? Or just pathetic?

That's not fair.

That's what Frankfurter's grandson would probably say. Only so much can be expected of someone in a moment of crisis. But you're the grandson of a Holocaust survivor whose siblings were raped and murdered, whose parents were shot while holding babies, whose grandparents were burned alive. What do you think would be fair to expect of Frankfurter?

But people really do have limits. Those limits are not their choice, and not their fault, no matter how harshly they are judged by history.

I don't know.

What's not to know?

Maybe we underestimate some limits, and overvalue some actions. The man who lifted the car off the trapped bicyclist exceeded his physical limits. But did he then go home and campaign for designated bike lanes and more traffic lights? Because bicyclists being killed by automobiles is a systemic problem, not one solved by singular acts of hysterical strength. Is it fair to ask if he did enough?

No, it's not fair, because he—

Did Karski do enough? You've made Frankfurter's lack of belief the subject of the story, but what about Karski's limits? He left Frankfurter without obtaining guarantees that a way had been decided upon to save the Jews. He didn't refuse food and drink and die a slow death in the justice's chambers. Is it fair for us to judge him? What about those whose lives, and children's lives, depended on his mission's success? Would it have been fair for them to judge him?

He disguised himself as a Jew—donned a yellow armband, wore a Star of David—in order to smuggle himself into the Warsaw Ghetto to document conditions. He infiltrated a Nazi death camp so that he could share its truths with the world. Yes, he did enough.

What about your grandmother?

I'm sure she would agree.

That's not what I mean. It feels cruel, even depraved, to ask if your grandmother did enough—

Don't.

—but did she do enough?

Stop.

She fled her shtetl because she knew she "had to do something." She knew. *Her sister followed her outside, gave your grandmother her only pair of shoes, and said, "You're so lucky to be leaving." Another way of saying, "Take me, too." Perhaps her sister was too young to make the trip and bringing her would have doomed them both. Perhaps what your grandmother believed at the time was far less than we assume she believed. But you daydream about going from house to house of her shtetl, grabbing the faces of those who would stay, screaming, "You have to do something!" Why didn't your grandmother grab their faces?*

Because that's too much to expect of someone.

I agree. It's far too much to expect of someone.

So why did you ask the question?

Because agreeing on what can't be asked of someone reminds us of just how much can. We might disagree about what Frankfurter could have done, but we agree that he could have done more than he did.

Yes.

Now imagine yourself eating a burger behind a closed door.

I feel ridiculous for—

Stop telling me what you feel. Tell me what you can do.

Of course I can eat fewer animal products. And of course my fear of inconsistency doesn't have to stop me from trying. Right now I feel really hopeful, but—

Stop telling me what you feel.

But that's because we're talking. Relative to historical traumas, and in the context of this kind of deep questioning, my need and ability to make small daily changes couldn't be more obvious. But I know what will happen: Time will pass, I'll lose my reference points, stop assessing my sacrifices on the scale of global calamity, and go back to comparing my life to itself. And no matter what I know and want, I'll find myself back where I started.

Don't do that.

Give up?

Emphasize hopefulness.

But it's motivating.

Sure, when you feel hopeful. But unless you're ignorant or delusional about climate change, most of the time you won't feel hopeful. So then what? If hope is your primary motivation, you'll be rowing a sailboat in the doldrums—staring at the limp sail, waiting for it to inflate and relieve what feels like an unfair burden. Noah's ark didn't have a sail, and neither does ours. Knowing that no one and nothing will help us makes the effort easier.

I'm not sure I have the energy to sustain this for the rest of my life. It's not just rowing—it's rowing against the current. I'm thinking about the thousands of breakfasts and lunches ahead of me, always having to give them thought, resist cravings, risk social tension.

Instead of imagining all the meals ahead of you, focus on the meal in front of you. Don't give up burgers for the rest of your life. Just order something different this one time. It's hard to change lifelong habits, but it's not that hard to change a meal. Over time, those meals become your new habits.

So why hasn't vegetarianism become any easier after thirty years? Why has it become harder? I crave meat more now than I have at any point since I became a vegetarian.

Is that so horrible?

It is when I act on the craving.

How many times have you eaten meat in the last decade?

I don't know. A couple dozen?

That's more than you suggested earlier in the book.

I was warming up.

Let's say you've eaten meat one hundred times.

I haven't.

Okay, so you've eaten it two hundred times. Of the last 10,950 meals, you've fallen short two hundred times. You're batting .982.

I haven't eaten it *nearly* two hundred times.

You ask why it hasn't become easier? I ask what makes it so easy.

Talking to you.

It's like that first suicide note, "Dispute with the Soul," except we have to make sure this conversation never ends.

I *want* it to end. I want to put this to rest, as I've put to rest the decisions not to murder people, steal, or litter. Some people convert to veganism and never go back. For some, it seems as simple as deciding not to be an arsonist—it's so obviously the right thing to do, it doesn't require any thought, much less struggle. But with food, I'm always finding myself back where I started.

You know that thing about sharks?

How they have to keep swimming or else they die?

Right. Except that's only true for a few species of sharks. Most sharks don't have to swim to breathe.

Wrong about ostriches, wrong about sharks.

But maybe you're not most sharks. Maybe some people will find it easy to eat fewer animal products, or go totally vegan, and they won't have to sustain a lifelong debate about it. You just have to accept that your mind and heart are not built that way. And I'd wager that most humans are not most sharks.

So what do I do when I find myself back at the start? Open a Word document and describe my patheticness to you?

No, you just have to acknowledge that finding yourself at the start is not a regression. To "find yourself" anywhere is a good thing—it implies self-awareness. If you were halfway through a marathon and suddenly entered the mind-set of having the full twenty-six miles still to go, you'd probably want to give up. But isn't the starting point further back, at the decision to run a marathon? And isn't that decision always made with resolve and some amount of joy? It's why people renew their vows—to revisit the foundation of the marriage. There's a balance to be struck, as we need to do certain things even if we don't feel like doing them—we can't wait for the right feelings. But at times, remembering why we care in the first place can be motivating. What is the foundational truth for you here?

What do you mean?

Is there an idea, maybe even a sentence, on which everything else builds?

Our planet is an animal farm.

Tell me about it.

I've already told you about it.

Tell me about it again. The retelling is as important as what is being told.

We've misunderstood what our planet is, and therefore misunderstood how to save it.

Tell me for real. We have time.

Our singular focus on fossil fuels leads us to visually represent the planetary crisis with smokestacks and polar bears. It's not that those things are unimportant, but as mascots for our crisis, they have given us the impression that our planet is a factory, and that the animals most relevant to climate change are wild and far away. Not only is that impression wrong—it is disastrously counterproductive. We will never address climate change, never save our home, until we acknowledge that our planet is an animal farm. That correction is my starting point.

I thought we were failing to address climate change because of denial?

That idea is an even more insidious kind of denial than the denial it refers to.

Tell me.

But you already know.

Tell me again.

It creates a dichotomy between those who accept the science and those who don't.

But that dichotomy is real, isn't it?

Real and trivial. The only dichotomy that matters is between those who act and those who don't. Frankfurter told Karski, "I am unable to believe what you told me." But imagine if it had gone differently. Imagine if he had said, "I believe you." Imagine if he had committed to doing everything he could to help save the Jews of Europe: convene a group of influential figures to hear what Karski had to say, urge Congress to open an official investigation into German atrocities, use his voice to publicly raise the urgent questions. And more.

That sounds good.

But then, after promising all that, and maybe even benefiting from the ethical glow it gave his image, he did nothing. No convening, no urging, no voicing. Worse, he refused even to participate in home-front efforts: he gorged on rationed foods, drove as much and as fast as he wanted, lived in the only house on the street whose lights stayed on through the night. Knowing that, would it matter how he answered a poll conducted in 1943 asking about attitudes toward the war in Europe?

At least Karski would have left their meeting with hope . . .

We dramatically overstate the role of science deniers, because it allows science acceptors to feel righteous without challenging us to act on the knowledge we accept. Only 14 percent of Americans deny climate change, which is a significantly lower percentage than who deny evolution, or that the earth orbits the sun. Sixty-nine percent of American voters—including the majority of Republicans—say that the United States should have remained in the Paris climate accord. The rhetoric and optics might have been co-opted by liberals, but there is nothing more conservative than conservation.

How do you explain all the people who don't deny that the planet is imperiled but aren't alarmed that the planet is imperiled?

I would probably call them stupid or evil, if I weren't among them.

You aren't alarmed?

I want to be, but I'm not. I say I am, but I'm not. And as the situation becomes ever more alarming, so does my ability to ignore the alarm.

How do you explain that?

I don't know.

Try.

Humans are singularly adaptable creatures.

That sounds like bullshit.

It is.

So try harder.

We—

Don't tell me about everyone. Tell me about you.

My strategy when I wrote "How to Prevent the Greatest Dying"—the most information-heavy pages in this book—was to be as attentive as possible to my own reactions, rather than to emulate the journalistic style of the articles and books I was reading as research, none of which—regardless of how thoughtful, well written, and urgently important they were—ultimately moved me to *do* anything. I was willing to trade comprehensiveness, even a kind of professionalism, for a form that motivated me.

Did it work?

I definitely convinced myself.

Isn't that good?

I convinced myself of what I was already convinced of, and didn't live any differently in response.

So maybe you're no better than your friend, after all? You wrote a book and don't believe it; he won't read a book because he does.

It's a shame that instead of having a minority of climate atheists, we have a majority of climate agnostics.

But you said most Americans wanted the United States to stay in the Paris accord?

They gave that answer in response to the question. I would have, too. It's too bad that such opinions are selfies and not carbon sinks.

So you're . . . not *hopeful?*

I'm not. I know too many smart and caring people—not advocacy narcissists, but good people who give their time, money, and energy to improve the world—who would never change how they eat, no matter how persuaded they were to do so.

These smart and caring people, how would they explain their unwillingness to eat differently?

They would never be asked to.

If they were?

They might say that animal agriculture is a system with serious flaws, but people have to eat, and animal products are cheaper now than they have ever been before.

And how would you respond to that?

I would say we have to eat, but we don't have to eat animal products—we are certainly healthier when plants make up the majority of our diet—and we clearly don't have to eat them in the historically unprecedented quantities that we currently do. But it's true that this is an issue of economic justice. We should talk about it as one, rather than use inequality as a way to avoid talking about inequality.

The richest 10 percent of the global population is responsible for half the carbon emissions; the poorest half is responsible for 10 percent. And those who are the least responsible for global warming are often the ones most punished by it. Consider Bangladesh, the country widely considered to be most vulnerable to climate change. An estimated six million Bangladeshis have already been displaced by environmental disasters like storm surges, tropical cyclones, droughts, and flooding, with millions more projected to become displaced in the coming years. Anticipated sea-level rises could submerge about one-third of the country, uprooting twenty-five to thirty million people.

It would be easy to hear that figure and not feel it. Every year, the *World Happiness Report* ranks the top fifty happiest countries in the world on the basis of how respondents score their lives, from "the best possible life" to "the worst possible life." In 2018, it ranked Finland, Norway, and Denmark as the three happiest countries in the world. When the rankings were released, they clogged NPR for a couple of days, and seemed to come up in every conversation. The combined population of Finland, Norway, and Denmark is approximately half of the number of anticipated Bangladeshi climate refugees. But those thirty million Bangladeshis who are threatened with the worst possible lives don't make for good radio.

Bangladesh has one of the smallest carbon footprints in the world, meaning it is least accountable for the damage that most afflicts it. The average Bangladeshi is responsible for 0.29 metric tons of CO_2e emissions per year, while the average Finn is responsible for about 38 times that: 11.15 metric tons. Bangladesh also happens to be one of the world's most vegetarian countries, where the average person consumes about nine pounds of meat per year. In 2018, the average Finn happily consumed that amount every eighteen days—and that doesn't include seafood.

Millions of Bangladeshis are paying for a resource-opulent lifestyle that they have never themselves enjoyed. Imagine if you had never touched a cigarette in your life but were forced to absorb the health tolls of a chain-smoker on the other side of the planet. Imagine if the smoker remained healthy and at the top of the happiness chart—smoking more cigarettes with each passing year, satisfying his addiction—while you suffered lung cancer.

Worldwide, more than 800 million people are underfed, and nearly 650 million are obese. More than 150 million children under the age of five are physically stunted because of malnutrition.

That's another figure that demands a pause. Imagine if everyone living in the United Kingdom and France were under five years old and without enough food to grow properly. Three million children under the age of five die of malnutrition *every year.* One and a half million children died in the Holocaust.

Land that could feed hungry populations is instead reserved for livestock that will feed overfed populations. When we think about food waste, we need to stop imagining half-eaten meals and instead focus on the waste involved in bringing food to the plate. It can require as many as twenty-six calories fed to an animal to produce just one calorie of meat. The UN's former special rapporteur on the right to food, Jean Ziegler, wrote that funneling one hundred million tons of grain and corn to biofuels is "a crime against humanity" in a world where almost a billion people are starving. We might call that crime "manslaughter." What he didn't mention is that every year, animal agriculture funnels more than seven times that amount of grain and corn—enough to feed every hungry person on the planet—to animals for affluent people to eat. We might call that crime "genocide."

So, no, factory farming does not "feed the world." Factory farming starves the world as it destroys it.

Presumably that would put that counterargument to bed.

There's a parallel argument that I often hear: advocating for plant-based diets is elitist.

Elitist how?

Not everyone has the resources to give up animal products. Twenty-three and a half million Americans live in food deserts, and nearly half of them have low incomes. No one would argue that the poor should pay for the behavior of the rich with flooding and famines and so on. But how can you ask them to pay for expensive foods?

And?

It is true that a healthy traditional diet is more expensive than an unhealthy one—about $550 more expensive over the course of a year. And everyone should, as a right, have access to affordable, healthy food. But a healthy vegetarian diet is, on average, about $750 *less* expensive per year than a healthy meat-based diet. (For perspective, the median income of a full-time American worker is $31,099.) In other words, it is about $200 cheaper per year to eat a healthy vegetarian diet than an unhealthy traditional diet. Not to mention the money saved by preventing diabetes, hypertension, heart disease, and cancer—all associated with the consumption of animal products. So, no, it is not elitist to suggest that a cheaper, healthier, more environmentally sustainable diet is better. But what does strike me as elitist? When someone uses the existence of people without access to healthy food as an excuse not to change, rather than as a motivation to help those people.

Any other counterarguments?

What about the millions of farmers who would be driven out of business?

What about them?

There are fewer American farmers today than there were during the Civil War, despite America's population being nearly eleven times greater. And if the ultimate dream of the animal-agriculture industrial complex is realized, there will soon be no farmers at all, because "farms" will be fully automated. I was happily surprised to find that animal farmers were some of the greatest allies of *Eating Animals*—they despise factory farming every bit as much as animal rights activists do, if for different reasons.

The planetary crisis will make it more difficult and more expensive to raise livestock, as droughts reduce crop yields, and extreme weather events—like hurricanes, wildfires, and heat waves—kill farm animals. Climate change is already causing losses to livestock farmers around the world. In the long run, transitions toward renewable energy, plant-based foods, and sustainable farming practices will create many more jobs than they end. These transitions will also save the planet, and what would it mean to save farmers without saving the planet?

What else?

Not all animal products are bad for the environment.

Which is bullshit because . . . ?

It isn't bullshit. It is entirely possible to raise a relatively small number of animals in environmentally sensitive ways. That's what farming used to be until the advent of the factory farm. It is also possible to smoke cigarettes without harming your health. A single cigarette will have no effect.

Yeah, but who smokes only one cigarette?

People who hate the experience, or people who know better and quit before they get addicted. It's the rarest of eaters who hates animal products. Most people, like me, love those foods. So we naturally want more. I know better but often find my cravings too powerful to override. Like most Americans, I grew up eating meat, dairy, and eggs, so I didn't have the chance to quit before I got addicted.

But generally, animal products are bad for the environment?

More than generally, and more than bad. According to the UN, animal agriculture is "one of the top two or three most significant contributors to the most serious environmental problems, at every scale from local to global . . . It should be a major policy focus when dealing with problems of land degradation, climate change and air pollution, water shortage and water pollution and loss of biodiversity. Livestock's contribution to environmental problems is on a massive scale."

So why bother mentioning that there's such a thing as a good farm?

Because it's extremely tempting to oversimplify this scientifically and psychologically complicated issue: cherry-pick convenient statistics, dismiss "illogical" feelings, ignore marginal cases. And when it's already so hard to take to heart that what we eat matters—when even smart and caring people seek holes they can escape through with their lifestyles intact—inaccuracies can feel like dishonesties.

That's another counterargument, by the way: the numbers are vague to the point of being untrustworthy. I've cited animal agriculture's contributions to global greenhouse gas emissions at 14.5 percent. I've also cited them at 51 percent. And the low estimate wasn't provided by Tyson Foods, and the high estimate wasn't provided by PETA. It is arguably the most important of all climate change statistics, and the high estimate is more than three-times the low estimate. If I can't be more precise than that, why should anyone trust what I'm saying?

Why should they?

I *can* be more precise. In the appendix, I present the methodology behind those numbers and explain why I think 51 percent is more accurate. But the systems in question are complex and linked, and quantifying them requires making significant assumptions. Even the most politically neutral scientists face that challenge.

Take the shift to electric cars. How do we factor in the relative cleanliness of the electric grid that powers the cars? In China, coal generates 47 percent of electricity; a shift to electric cars would be a climate-change catastrophe. How do we consider the fact that it takes about twice the amount of energy to produce an electric car as it does a conventional one? And what about the other forms of environmental damage, like the mining of rare minerals for batteries, an energy-intensive process that can make use of only about 0.2 percent of what is pulled from the ground—with the other 99.8 percent (now toxically contaminated) being thrown back as untreated pollution?

It is dangerous to pretend that we know more than we do. But it is even more dangerous to pretend that we know less. The difference between 14.5 percent and 51 percent is enormous, but even the low end makes absolutely clear that if we want to stop climate change, we cannot ignore the contributions of animal products.

Frankfurter asked Karski about the height of the Warsaw Ghetto's wall. If Karski had answered that it was between eight and twenty-five feet, would that have made any difference? To the Jews unable to scale it? To Frankfurter as he considered their fate? To us as we judge Frankfurter?

But without knowing the height of the wall, we can't plan how to overcome it.

Different studies suggest different dietary changes in response to climate change, but the ballpark is pretty clear. The most comprehensive assessment of the livestock industry's environmental impact was published in *Nature* in October 2018. After analyzing food-production systems from every country around the world, the authors concluded that while undernourished people living in poverty across the globe could actually eat a little more meat and dairy, the average world citizen needs to shift to a plant-based diet in order to prevent catastrophic, irreversible environmental damage. The average U.S. and U.K. citizen must consume 90 percent less beef and 60 percent less dairy.

How would anyone keep track of that?

No animal products for breakfast or lunch. It might not amount to precisely the reductions that are asked for, but it's just about right, and easy to remember.

And easy to do?

Depends on the shark. It would be both disingenuous and counterproductive to pretend that eating only plant-based foods before dinner won't require some adjusting. But I bet that if most people think back over their favorite meals of the past few years—the meals that brought them the most culinary and social pleasure, that meant the most culturally or religiously—virtually all of them would be dinners.

And we have to acknowledge that change is inevitable. We can choose to make changes, or we can be subject to other changes—mass migration, disease, armed conflict, a greatly diminished quality of life—but there is no future without change. The luxury of choosing which changes we prefer has an expiration date.

For you?

What?

Has the change been easy for you?

I've set myself the deadline of finishing this book to give up dairy and eggs.

You're joking.

I'm not.

You mean you haven't been consistent with it yet?

I haven't yet tried.

How the hell do you explain that*?*

With the only counterargument that leaves me stumped: this is a fantasy. It is a scientifically sound fantasy, an ethical fantasy, an irrefutable fantasy. But a fantasy. Large numbers of people are not going to change how they eat, certainly not in the required time. Clinging to a fantasy is every bit as dangerous as dismissing a viable plan.

And how would you respond to that?

Being living proof of their argument, I would have a very difficult time.

Try.

The truth is, I'm not hopeful.

Good. Now tell me how the fantasy could be a viable plan.

It's hard to imagine.

Even if it's the longest of shots.

If it happens, there won't be any one thing that makes it happen. Doing what needs to be done will involve invention (like creating veggie burgers that are indistinguishable from beef hamburgers), and legislation (like adjusting farm subsidies and holding animal agriculture responsible for its environmental destruction), and bottom-up advocacy (like college students demanding their cafeterias not serve animal products before dinner), and top-down advocacy (like celebrities spreading the message that we cannot save the planet without changing how we eat), and—

No one will cure climate change? Everyone will cure climate change?

Exactly.

Help me to see how.

Honestly, I can't see it.

You're hopeless.

I'm realistic.

And even if you think I already know, remind me why it is realistic to be hopeless?

You're joking?

Say it.

Because of the destruction we've already wreaked—destruction that either has to be undone or cannot be undone. Because of the fact that within a single year, loggers destroyed an area five times the size of London in the Amazon, an ecosystem that takes four thousand years to regenerate. Because of how difficult it will be to reverse a way of being with 7.5 billion people putting their weight on the gas pedal. Because American CO_2e emissions increased by 3.4 percent in 2018. Because of the imprecision of math that depends on precision—the difference of half a degree could be the difference that makes all the difference. Because of the justified desire of developing countries to become like the countries most responsible for climate change. Because as it gets hotter, more air-conditioning will be used, which will expel more greenhouse gases. Because of the thousands of other positive feedback loops. Because of the 2017 discovery that methane emissions from cattle are at least 11 percent higher than previously thought and the 2018 discovery that oceans are warming 40 percent faster than previously thought. Because many of the people most affected by climate change (and best able to testify to its horrors) don't have the means to share their testimonies and shake our collective conscience. Because the interests motivated against solving the problem are more powerful, driven, and clever than the interests motivated to solve it. Because in the next thirty years, the human population is expected to increase by 2.3 billion, and global income is expected to triple, meaning many more people will be able to eat diets rich in animal products. Because of the seeming impossibility of cooperation across countries and within countries. Because there's a very good chance it's already too late to avoid runaway climate change. Because—

I get it.

Because of human nature: people like me, who should care and should be motivated and should make big changes, find it almost impossible to make small sacrifices for profound future benefit. Because—

Enough.

Because I haven't even tried.

I don't know.

What's not to know?

Why are we still talking?

What do you mean?

You just told me you haven't even tried, and yet we're still talking.

So?

Remember "Dispute with the Soul of One Who Is Tired of Life"?

I'm not writing a suicide note.

That's my point. And my stubborn hope.

I thought you were anti hope.

I'm anti stolen hope.

And the price of hope is action.

And there is one action that gives me hope.

Giving up animal products?

No.

You've lost me.

I haven't. Not yet. We're still talking, so you haven't yet lost yourself.

What are you saying?

Suicide notes end. We're still swimming. This is what it looks like to try. Are you tired?

Of this conversation? Yes.

Of life.

No.

"Dispute with the Soul of One Who Is Not Yet Tired of Life." But it's wrong to assume that the soul is what we appeal to with the momentous questions in the momentous moments: How should I live? Whom should I love? What is the purpose? It's the soul that asks *the questions, not answers them. The soul is no more "over there" than the causes and solutions to climate change. Even worse, we are tragically confused about what is momentous.*

Confused how?

We ask the soul, "Are you hopeful?" The soul asks us, "What's for lunch?"

Mr. Karski.

What about him?

Mr. Karski, a man like me talking to a man like you must be totally frank.

I'm Karski?

I must say I am unable to believe what you told me.

You think I have been lying to you?

I didn't say that you were lying. I said I am unable to believe you. My mind, my heart, they are made in such a way that I cannot accept it.

Made by whom?

I'm sorry, but I have an urgent matter to attend to.

Mr. Karski.

. . . Yes?

A man like me talking to a man like you must be totally frank.

You think I have been lying to you?

I don't know.

What's not to know?

How tall is the ice shelf?

Two hundred feet.

That doesn't sound so bad.

One hundred feet.

I don't know.

Mr. Karski.

Yes.

I want to believe you.

Is scale the problem? That the immensity of the tragedy forces it into abstraction? Because I was lying earlier.

I didn't say you were lying.

Only a few thousand children are dying of malnutrition. Now will you do something to save them?

That's not the problem.

Is it distance? I made it seem far away so that you wouldn't be frightened, but the Supreme Court will be underwater.

Distance isn't the problem.

I am trapped beneath a car.

Excuse me?

I need you to lift it off me.

There is no car.

Why won't you save my life?

Because it clearly doesn't need to be saved.

So why won't you save the lives that clearly need to be saved?

Because I am also trapped under a car.

Mr. Karski, a man like me talking to a man like you must be totally frank.

Who cares about frankness now?

Mr. Karski. I've given you my time, heard you out, told you my position. Now you must leave.

I accept that you don't believe me. I rarely believe myself. I don't need you to believe me.

Go!

I need you to act.

Next time, I won't even let you into this room.

Next time?

Next time I replay this conversation in my mind.

The ice shelf could fit under your door.

Is that tall?

Why don't you have any children, Mr. Karski?

We didn't want any.

Why didn't you want any?

We were happy as we were.

Is it because you are also doomed to forever replay this conversation in your mind?

Why don't you have any children, Justice Frankfurter?

Is that any of your business?

Why does my question make you defensive?

Marion suffered a great deal. She was frail. It would have been too much.

I am unable to believe you.

You think I am lying?

I didn't say that you were lying. I think you cannot admit, even to yourself, that the prospect of a child's judgment prevented you from having children.

Mr. Karski.

Your mind, your heart.

Yes. They are made in such a way that I cannot accept what you have told me. Not because they are deficient. Because they function. If I were to accept what you have said, I would go mad.

You would act.

I would know that no action would be enough.

You could refuse food and drink, die a slow death while the world looked on.

That wouldn't be enough.

You could convene a group of influential figures to hear what I have to say, urge Congress to open an official investigation into climate atrocities, use your voice to publicly raise the urgent questions.

That wouldn't be enough.

After I leave, you could eat a different kind of lunch than you otherwise would have chosen.

I don't know.

Mr. Karski.

I lied about the height of the ice shelf.

I didn't say that you were lying.

But I was.

So how tall is it?

This tall.

As tall as the walls of this room?

As tall as the page on which these words are printed. Not *as tall as*. This page *is* the wall. The other side of it.

I don't understand.

No matter how far away your obligations feel, no matter the height or thickness of the ice that separates you from them, they are only on the other side. Right there. Right here.

I don't know.

Mr. Karski.

What's not to know!

I don't know.

I must go now.

Mr. Karski!

The wall is melting, and I have an urgent matter to attend to.

More urgent than this?

I need to go back, and tell them what happened here, and implore them to save you.

Save *me?*

They must do more: die in greater numbers, more quickly, more grotesquely. They must do their part, create a spectacle of suffering that demands a response.

Keep talking to me.

What good would that do? Your mind, your heart, they are made in such a way that you cannot accept what I say.

But they are always being made.

I worry.

That I won't change?

I worry that they won't believe your disbelief.

V. MORE LIFE

Finite Resources

Returning from work one afternoon, my grandfather was stopped on the outskirts of his Polish village by a friend, who told him that everyone had been murdered and that he had to flee. "Everyone" included my grandfather's wife and baby daughter. He wanted to turn himself in to the Nazis, but his friend physically restrained him and forced my grandfather to survive. After several years of running and hiding, exercising hysterical resourcefulness to evade the Germans, he met my grandmother, and they moved to Lodz, where they lived in an empty home whose previous occupants had been murdered.

Resourcefulness was the only quality that I heard attributed to my grandfather until a few years ago. He ran the black market of the displaced persons camp where he, my grandmother, and my mother spent their final months in Europe; traded currencies and precious metals; forged documents; hid his money in the carved-out heels of his shoes. In 1949, he and his young family boarded a ship to America with a suitcase that held ten thousand dollars in cash—today's equivalent of more than one hundred thousand dollars. (They had more money than the

American relatives who were taking them in.) Speaking little English, and unacquainted with American culture or business, he bought a series of small grocery stores, managed them, and then sold them for a profit. Such stories about him—and all stories about him were such stories—filled me with pride, as well as some embarrassment about my own relative lack of ingenuity.

Around the time my mother was six, my grandfather said he was going downstairs to get the store ready—they often lived above the groceries they owned—and hanged himself from one of the air-conditioning units. Just when the threats appeared to have cleared, his resourcefulness, his ability to survive anything, reached its limit. He was forty-four.

I didn't know about my grandfather's suicide until a series of somewhat accidental discoveries a decade ago. Clearly an earlier confrontation with the truth wouldn't have changed any of the facts, but it might have spared my family unnecessary shame and the guilt that incubated in the silence.

To some extent, we all knew what we didn't know. Or we knew but didn't believe, and in that way didn't know.

My mother recently told me that she remembers when her father tucked her in for the last time. "He kept kissing me and telling me that he loved me in Yiddish."

She believes that while he suffered from clinical depression, his suicide was triggered by a failed business venture, which would encumber the family with extreme debt—the shame of leaving his wife and children without enough resources compelled him to take away their greatest resource.

Maybe resourcefulness really did define him so completely. Or maybe that description was a powerful act of repression, avoiding a truth by asserting its opposite. Maybe "resourceful," counterintuitively, is a description of someone who survives with very few resources. Or maybe it is a description that means nothing at all, given to someone who was hardly known—another way of saying "he lived."

America is also known for its "resourcefulness," due to both its innovation and its consumption. And although there is a temptation to describe the planetary crisis in apocalyptic terms, imagining total human extinction, the truth is that many of us who live in high-income nations with varied landscapes and sophisticated technology will survive our climate suicide. But we will suffer permanent injuries. When Kevin Hines jumped off the Golden Gate Bridge, he shattered two vertebrae—most survivors sustain broken bones and punctured organs. We will be displaced by extreme weather, our coasts will become unlivable, and our economy will crash. Armed conflicts will erupt, food prices will soar, water will be rationed, pollution-related illnesses will skyrocket, mosquitoes will invade. And our psychologies will be changed by the traumas: being separated from our families in extreme weather events, leaving aging parents behind in places debilitated by drought or flooding so that children might have less arduous lives, competing for resources in ways more explicit and less civilized than we ever have before.

If we regard American apathy toward climate change as a kind of suicide, our suicide is made grislier by the fact that we aren't primarily the ones to die from it. Most of the populations that are already dying from climate change, and the populations that climate change will kill in the future, reside in places with minimal carbon footprints, places like Bangladesh, Haiti, Zimbabwe, Fiji, Sri Lanka, Vietnam, and India. They will not die for lack of resourcefulness.

At this moment in the environmental movement, we can

jump from the bridge, or we can cross it. We can allow the fear that it's too late or too difficult to ensure resources for future generations to incapacitate us, or we can allow those fears to capacitate us. We are their—and our—most valuable resource.

Earth is about 4.5 billion years old. I am nearly my grandfather's age when he killed himself. In the sense of facing the choice of whether to live, we all are his age.

The Flood and the Ark

One of the largest mass suicides in history is one of my culture's seminal stories. Around 72 C.E., troops from the Roman Empire laid siege to the mountaintop Jewish community of Masada. For at least a month and a half, the greatly outnumbered Jews held off their Roman attackers. But when it became clear that the battle was lost, they committed suicide to avoid capture. As taking one's own life is prohibited by Jewish law, the citizens of Masada drew lots and killed each other in turn, until there was only one Jew remaining—the only one then required to break Jewish law. The only one to die from his suicide.

I visited Masada as a child and was encouraged to see the collective suicide as a symbol of Jewish resistance: a heroic alternative to submission. But it struck me, even then, as fanatical. Why not try to negotiate a surrender? Why not just pretend to convert? Why not live to fight another day, or at least live to live another day?

The Masada suicide was also at odds with the heroism I was taught to revere with other foundational stories, like that of the Jews of the Warsaw Ghetto in World War II—those whose fate Jan Karski came to America to share. In a situation no less dire than that of the Jews of Masada, the Warsaw Ghetto Jews battled to the end. They dug underground bunkers and tunnels, built passages across rooftops, stole a tiny arsenal of guns, made their own crude weapons, initiated an armed resistance, and fought until no more fighting was possible.

Much of what is known of the Warsaw Ghetto comes by way of the Ringelblum Archive, a collection of testimonies, artifacts, and documents secretly collected by a team of ghetto Jews, led by the historian Emanuel Ringelblum. More than thirty-five thousand pages were placed in milk cans and buried for future discovery. As one of the documents, written by a nineteen-year-old in 1942, attests: "What we were unable to cry and shriek out to the world we buried in the ground . . . I would love to see the moment in which the great treasure will be dug up and scream the truth at the world . . . May the treasure fall into good hands, may it last into better times, may it alarm and alert the world to what happened."

What is known of the Masada mass suicide comes by way of Flavius Josephus. There is ample archaeological evidence of a Jewish community at Masada, but little reason to believe the historical accuracy of Josephus's story. The myth of the collective suicide has been perpetuated and spread because there have been such strong incentives to keep those deaths alive. A tiny, new country, surrounded by neighbors that dwarf it and want to destroy it, needs others to believe in its unconditional refusal to surrender. And it needs to believe in it as well.

•

Carved into the rock of a permafrost mountain in Norway, 130 meters above sea level, is the Svalbard Global Seed Vault, the world's largest collection of agricultural biodiversity. In the event of a total agricultural collapse, the global seed vault could supply food security.

The structure was built to withstand the test of time, extreme weather, and human attacks. But in 2017—the world's hottest year on record—unusual melting and rain flooded the tunnel entrance. Because the vault is kept at minus 18 degrees Celsius (minus 0.4 degrees Fahrenheit), the water froze and did not reach the seeds. Now Norway plans to spend about $12.7 million to implement even more protective measures. But the episode proved that even a structure designed to endure "the challenge of natural or man-made disasters" may not be able to endure a man-made natural disaster.

Another effort, called the Frozen Ark Project, strives "to facilitate and promote the conservation of tissue, cells and DNA from endangered animals." Separately, Moscow State University recently received Russia's largest ever scientific grant to create a DNA bank, dubbed "Noah's Ark," whose goal is to include genetic material from every living and extinct species of organism.

The Masada story and the Ringelblum Archive are seed vaults that we can draw upon. So is my daydream of going back in time and warning the Jews of my grandmother's village: "You have to do something!" So is Karski's successful effort to inform Frankfurter but failed effort to move him. In times of unprecedented threat, we can reach into history for help. We can also reach into the future. In the final lines of *An Inconvenient Truth*, Al Gore says, "Future generations may well have occasion to ask themselves, 'What were our parents thinking? Why didn't they wake up when they had a chance?' We have to hear that question from them now."

We can uncover testimonies from the past, listen to testimonies from the present, and imagine testimonies from the future. But it is not enough to be convinced by them. We need to be converted.

•

The Noah invoked by the DNA-bank projects was the first person born in the world after Adam's death—the first person who could have no direct contact with a living memory of Eden. He was the first person to enter a world in which natural death was present, the first to age knowing that he must die.

It is written that "Noah was a righteous man, blameless in his time." Why "blameless in his time" and not simply "righteous"? Because righteousness and blame are contextual. Being a good person at Normandy on June 6, 1944, is not the same as being a good person in a grocery store in 2019. The Warsaw Ghetto demanded something different from what Superstorm Sandy did. Eating blamelessly two generations ago is not the same as eating blamelessly in the age of the factory farm. Just as a situation can inspire hysterical strength, it can also inspire, and require, an unprecedented ethical response. The something we have to do must respond to the something that needs to be done.

Noah is described as *ish ha'adama*, a "man of the earth"—an ironic, or perhaps perfectly fitting, title for someone most strongly associated with a deluge. About one hundred years pass between God instructing Noah to build the ark and God enacting the flood. A century might seem like a long time, but even in the context of a biblical story, it is remarkable that a man and his sons (with no modern tools, no electricity, no Home Depot) were able to build a structure large enough to save two of every kind of animal so quickly.

But a century is an almost impossibly long period to have to sustain belief. Imagine what those years must have been like for Noah—every day being dismissed as crazy, every day committing his full identity (his labors, his resources, his purpose) to something that could not be proved. The further time separated him from God's instruction—the more *over there* the command felt—the more difficult it must have been to maintain the necessary conviction. It must have required a constant internal dialogue, and a steady supply of apologies. Would civilians participate in blackouts for a war one hundred years in the future?

Yet Noah was luckier than we are. We have far less than a century to construct our ark—we have, perhaps, a decade to make the changes we haven't yet found a way to honestly dispute, with others or with ourselves. And unlike Noah, we have to do it without belief. Without instructions from above, we not only have to motivate ourselves to act, we also have to choose what kind of ark to construct. Our ark could be a spacecraft for colonizing Mars. It could be a seed bank to start over after the collapse of plant life or a DNA bank to start over after the collapse of animal life. It could be an act of collective suicide. Or it could be a wave of collective action.

After the floodwaters receded, God presented the rainbow as a symbol of His covenant with all creation never again to destroy the earth: this planet will be our only home. "My bow I have set in the clouds to be a sign of the covenant between Me and the Earth, and so, when I send clouds over the Earth, the bow will appear in the cloud. Then I will remember My covenant, between Me and you and every living creature of all flesh, and the waters will no more become a Flood to destroy all flesh. And the bow shall be in the cloud and I will see it, to remember the everlasting covenant between God and all living creatures, all flesh that is on the Earth."

He uses the word "remember" twice. It is strange that an all-powerful being would need an aid to remember not to eradicate His most important creation. The God of the Torah is forgetful, and requires reminders—the groaning of slaves in Egypt, symbols of His covenants—and He makes clear that the reminder is *for Him*. But this is no "note to self" on the bedside pad of paper. God's reminder is dramatic and public—literally written across the sky. So whatever the intent, the rainbow is also a memory aid for Noah. For humanity. We are reminded of what God did to us, and for us, and what God promised. But more, the rainbow reminds us of the possibility of destruction, which reminds us of something that seems so essential, it shouldn't require any reminder, but because it's so essential, requires a reminder more than anything else does: we don't want to be destroyed.

Globally, more people die of suicide than war, murder, and natural disasters combined. We are more likely to kill ourselves than we are to be killed and, in that sense, ought to fear ourselves more than we fear others. A rainbow is also a rope: it can be thrown to a drowning person, or it can be tied into a noose.

No one who isn't us is going to destroy Earth, and no one who isn't us is going to save it. The most hopeless conditions can inspire the most hopeful actions. We have found ways to restore life on Earth in the event of a total collapse because we have found ways to cause a total collapse of life on Earth. We are the flood, and we are the ark.

That Is the Question

On the morning of April 14, 2018, the civil rights attorney David Buckel entered a section of Brooklyn's Prospect Park that I have entered thousands of times. When I lived in the neighborhood, it's where I would often go to walk my dog, play with my children, or simply gather my thoughts. At 5:55 a.m., Buckel sent an e-mail to news outlets explaining the decision he was about to make. He then doused himself in gasoline and set himself on fire.

According to his husband and friends, he had not been depressed. And he had enough presence of mind to leave at least four separate messages explaining his act. The briefest of these notes was handwritten: "I am David Buckel and I just killed myself by fire as a protest suicide."

A second note was found wrapped in a garbage bag in a shopping cart nearby. It read: "Pollution ravages our planet, oozing inhabitability via air, soil, water and weather. Our present grows more desperate, our future needs more than we've been doing."

Buckel was a civil rights lawyer who had every reason to believe that progress was more than fantasy. He was a nationally recognized pioneer for gay and transgender rights. Gay marriage was legalized in Buckel's adulthood, thanks in part to his efforts. In an atmosphere of apathy and resignation, he seemed hopeful and motivated. Those who have characterized his suicide as an act of defeatism ignore the fact that his death was explicitly a protest. And there is no action more dependent on the belief that things could be different than a protest. "Honorable purpose in life invites honorable purpose in death," Buckel said in his suicide note.

•

Three months later, *The New York Times* ran the essay "Raising My Child in a Doomed World." Half a dozen friends sent it to me. On the first read, I found it poignant. Its author was Roy Scranton, the same man who wrote "Learning How to Die in the Anthropocene." Scranton describes the powerful mixture of emotions he felt upon the birth of his child: "I cried two times when my daughter was born." First came tears of joy, then sadness: "My partner and I had, in our selfishness, doomed our daughter to life on a dystopian planet, and I could see no way to shield her from the future."

I was grateful for another addition to the conversation about the planetary crisis. Scranton is not only thoughtful but passionate, informed, and a damned good writer. He voiced something that I have often felt as a parent. And it's no coincidence that so many people forwarded it to me, and that all of them were parents.

In this essay (and others), Scranton addresses the environmental crisis with the kind of philosophical rigor that is lacking from the present dialogue—a kind of thinking that we desperately need in order to comprehend our crisis. As David Wallace-Wells observes in his "Uninhabitable Earth" article, "We have not developed much of a religion of meaning around climate change that might comfort us, or give us purpose, in the face of possible annihilation." Scranton advances a religion here, but it doesn't give us purpose in the face of annihilation. As I reread the essay, I felt frustration, even anger. The longer I spent with it, the more it read as a kind of suicide note.

When considering the "ethics of living in a carbon-fueled consumer society," Scranton notes that many advocate for living more responsibly. "Take the widely cited 2017 research letter by the geographer Seth Wynes and the environmental scientist Kimberly Nicholas, which argues that the most effective steps any of us can take to decrease carbon emissions are to eat a plant-based diet, avoid flying, live car free and have one fewer child." (He is referring to a paper I cited earlier, "The Climate Mitigation Gap: Education and Government Recommendations Miss the Most Effective Individual Actions," which argues that most of the efforts to curb climate change that are taught and recommended are relatively insignificant.) "The main problem with this proposal," he continues, "isn't with the ideas of teaching thrift, flying less or going vegetarian, which are all well and good, but rather with the social model such recommendations rely on: the idea that we can save the world through individual consumer choices. We cannot."

Why not?

Because the world is a "complex, recursive dynamic" with "internal and external drivers."

I'm not exactly sure what that means, but however complex the world is, people still recycle, protest, vote, pick up litter, support ethical brands, donate blood, intervene when someone appears in danger, challenge racist remarks, and get out of the way of ambulances. These actions are not merely good for the individual health of the actor, but essential for the health of society: actions are witnessed and replicated.

In their book *Connected: The Surprising Power of Our Social Networks and How They Shape Our Lives*, Nicholas A. Christakis and James H. Fowler call social networks "a kind of human superorganism." They write, "We discovered that if your friend's friend's friend gained weight, you gained weight. We discovered that if your friend's friend's friend stopped smoking, you stopped smoking. And we discovered that if your friend's friend's friend became happy, you became happy." Although we often refer to obesity as an epidemic, it is rarely described as contagious. But Christakis and Fowler illustrate that—like smoking and the rejection of smoking, sexual misconduct and the rejection of sexual misconduct—obesity is a trend:

> In a surprising regularity that, as we have discovered, appears in many network phenomena, the clustering obeyed our Three Degrees of Influence Rule: the average obese person was more likely to have friends, friends of friends, and friends of friends of friends who were obese than would be expected due to chance alone. The average nonobese person was, similarly, more likely to have nonobese contacts up to three degrees of separation. Beyond three degrees, the clustering stopped. In fact, people seem to occupy niches within the network where weight gain or weight loss becomes a kind of local standard.

When it comes to health, this research suggests that individual behavior is much more impactful than federal dietary guidelines, which most Americans do not meet. While structures matter—food deserts, subsidies, and unhealthy cafeterias undeniably influence diet—the most contagious standards are the ones that we model.

We aren't powerless within our "complex, recursive dynamic" with "internal and external drivers"—we *are* the internal drivers. Yes, there are powerful systems—capitalism, factory farming, the fossil fuel industrial complex—that are difficult to disassemble. No one motorist can cause a traffic jam. But no traffic jam can exist without individual motorists. We are stuck in traffic because we *are* the traffic. The ways we live our lives, the actions we take and don't take, can feed the systemic problems, and they can also change them: lawsuits brought by individuals changed the Boy Scouts, the testimonies of individuals initiated the #MeToo movement, individuals participating in the March on Washington for Jobs and Freedom paved the way for the Civil Rights Act of 1964 and the Voting Rights Act of 1965. Just as Rosa Parks helped desegregate public transportation, just as Elvis helped prevent polio.

Scranton writes: "We are not free to choose how we live any more than we are free to break the laws of physics. We choose from possible options, not ex nihilo."

Yes, there are constraints on our actions, conventions and structural injustices that set the parameters of possibility. Our free will is not omnipotent—we can't do whatever we want. But, as Scranton says, we are free to choose from possible options. And one of our options is to make environmentally conscientious choices. It doesn't require breaking the laws of physics—or even electing a green president—to select something plant-based from a menu or at the grocery store. And although it may be a neoliberal myth that individual decisions have ultimate power, it is a defeatist myth that individual decisions have no power at all.

Both macro and micro actions have power, and when it comes to mitigating our planetary destruction, it is unethical to dismiss either, or to proclaim that because the large cannot be achieved, the small should not be attempted.

We need structural change, yes—we need a global shift away from fossil fuels and toward renewable energy. We need to enforce something akin to a carbon tax, mandate environmental-impact labels for products, replace plastic with sustainable solutions, and build walkable cities. We need structures to nudge us toward choices we already want to make. We need to ethically address the West's relationship to the Global South. We might even need a political revolution. These changes will require shifts that individuals alone cannot realize. But putting aside the fact that collective revolutions are made up of individuals, led by individuals, and reinforced by thousands of individual revolutions, we would have no chance of achieving our goal of limiting environmental destruction if individuals don't make the very individual decision to eat differently. Of course it's true that one person deciding to eat a plant-based diet will not change the world, but of course it's true that the sum of millions of such decisions will.

In response to the lifestyle changes proposed by Wynes and Nicholas, Scranton writes:

> To take [their] recommendations to heart would mean cutting oneself off from modern life. It would mean choosing a hermetic, isolated existence and giving up any deep connection to the future. Indeed, taking Wynes and Nicholas's argument seriously would mean acknowledging that the only truly moral response to global climate change is to commit suicide. There is simply

> no more effective way to shrink your carbon footprint. Once you're dead, you won't use any more electricity, you won't eat any more meat, you won't burn any more gasoline, and you certainly won't have any more children. If you really want to save the planet, you should die.

This is an extreme leap. Imagine yourself choosing not to eat animal products before dinner, choosing to take two fewer flights a year. Putting aside whether that would be possible for you, does it seem like "a hermetic, isolating existence"? Or does it seem like a reasonable adjustment? It's true that making decisions on behalf of the planet's health will cut us off from unbridled hedonism, but is this how we define "modern life"? If so, it should be a relief to cut ourselves off from it. It is only by making such decisions, such adjustments, that we will *ensure* a "deep connection to the future."

There is no more effective way to shrink one's carbon footprint than to die, but that suggests that everyone's carbon footprint is independent. Unless you buy and eat your food in secret, you don't eat alone. Our food choices are social contagions, always influencing others around us—supermarkets track each item sold, restaurants adjust their menus to demand, food services look at what gets thrown away, and we order "what she's having." We eat as families, communities, nations, and increasingly as a planet. Individual consumer choices can activate a "complex, recursive dynamic"—collective action—that is generative, not paralyzing. While the act of suicide can influence others, it is a final influence. We couldn't stop our eating from radiating influence even if we wanted to.

Even more important is the question of what we're trying to save. "If you really want to save the planet, you should die,"

Scranton writes. But the planet *isn't* what we want to save. We want to save life on the planet—plant life, animal life, and human life. Accepting the necessary violence of our existence is the first step to minimizing it: we must consume resources in order to survive. This would remain true in any political utopia. But plenty of species, including humans, have managed to live in harmony with nature, and they do not do so by committing suicide. They do so by taking less than the planet is able to produce and nurturing ecosystems. They do so by living as though we have only one Earth, not four. By treating the planet like our only home.

Scranton proceeds to describe the suicide of David Buckel, concluding that his "self-sacrifice takes the logic of personal choice to its ultimate end."

I do not condone Buckel's suicide, or any suicide. But it is important to remember that he did not kill himself to cap his carbon footprint. His self-immolation, in the tradition of Buddhist monks who publicly set fire to themselves to protest the Vietnam War, was explicitly designed to be witnessed: to burn into the public consciousness, to incite change. It weaponized an act of self-destruction to remind us that we don't want to self-destruct.

> The real choice we all face is not what to buy, whether to fly or whether to have children but whether we are willing to commit to living ethically in a broken world, a world in which human beings are dependent for collective survival on a kind of ecological grace.

What does it mean to live ethically if not to make ethical choices? Among these choices are what to buy, whether to fly, and how many children to have. What is ecological grace if not the sum of daily, hourly decisions to take less than one's hands can hold, to eat other than what our stomachs most want, to create limits for ourselves so that we all might be able to share in what's left?

> I can't protect my daughter from the future and I can't even promise her a better life. All I can do is teach her: teach her how to care, how to be kind and how to live within the limits of nature's grace. I can teach her to be tough but resilient, adaptable and prudent, because she's going to have to struggle for what she needs. But I also need to teach her to fight for what's right, because none of us is in this alone. I need to teach her that all things die, even her and me and her mother and the world we know, but that coming to terms with this difficult truth is the beginning of wisdom.

This is not the beginning of wisdom. This is the end of resignation.

Who cares if his daughter cares? Her grandchildren won't. What will matter most to them is not whether she was kind, or tough but resilient, or adaptable and prudent. What will matter most to them is whether she did what was necessary. The future does not depend on our feelings, and to a great extent, it depends on us getting over our feelings.

Scranton is right that none of us is in this alone. Why not teach his daughter that if she eats differently, and convinces others to do so, she—*they*—*we*—could participate in saving the planet? Rather than preparing her to "struggle for what she needs," how about struggling for what we all need? Eating less meat, flying less, driving less, having fewer children—these choices are struggles. If they weren't, we'd have made them long ago. I have not yet managed to cut dairy and eggs out of my diet. If I were any other kind of animal, my obligations would end with my desire. But I am a human, and that is where my obligations begin. The decision to fight for what's right requires us to cut ourselves off from what's wrong.

I have never met Roy Scranton, and I've never met his daughter, but I have obligations to them, just as they have obligations to my family. Just as Americans have obligations to Bangladeshis. Just as affluent suburbanites have obligations to those living in urban heat islands and food deserts. Just as people alive today have obligations to future generations.

•

I agree with Scranton that we can't properly conceptualize the environmental crisis—we certainly can't be alarmed by it—until we acknowledge its ability to kill us. Because we created it, this means we have to acknowledge our ability to kill ourselves. We have to be aware of the death that surrounds us, even when it hasn't yet happened, even when it is easy to miss, even when our suicide will kill others first.

A few months ago, a man committed suicide in his car only a few blocks from my office at New York University. Despite our age of sharing and voyeurism, and despite New York overflowing with pedestrians and surveillance cameras, his dead body remained unnoticed in the car for a week. A real estate agent with a nearby office parked his motorcycle in front of the vehicle. He couldn't believe that a body was inside, or that it was there for as long as it was. Traffic officers who write tickets on days of alternate-side parking will often ignore cars standing on the wrong side, so long as there is a driver at the wheel. Presumably a number of officers saw the body but assumed the man was living. A child complained of a terrible smell while passing the car and vomited on the sidewalk. His mother didn't notice anything. Someone walking his dog noticed a figure in the car and thought it was a napping Uber driver. When the body was still there, two days later, he called 911.

There are only two reactions to climate change: resignation or resistance. We can submit to death, or we can use the prospect of death to emphasize life. We will never know what the author of "Dispute with the Soul of One Who Is Tired of Life" chose. We do not yet know what we will choose.

It is horrible to imagine coming upon David Buckel's charred corpse. It feels yet more horrible to imagine passing a dead body many times without noticing it. But there is something even worse: not to notice that we are alive.

Four days after Buckel's suicide, one of the joggers who came upon his body wrote a beautiful short essay, reflecting on literal and metaphorical running. But it was her description of the park that morning, her first minutes before coming upon Buckel, that remains with me. She had just returned from a trip abroad and was eager to get some exercise. "The birds chirped, the sun shone, and as I made my way through the tree-lined paths, I felt bathed in a sense of joy of being back home and alive."

If all goes according to nature's plan, Buckel's daughter, Scranton's daughter, and my sons will live on the planet without their parents. I hope they will feel bathed in a sense of joy of being home and alive. I hope that their parents, in their own ways, to the best of their judgments and abilities, will have done what they had to do to allow for that. I hope we will have taught them—not only with our words but with our choices—the difference between running toward death, running away from death, and running toward life.

After Us

I am sitting at my grandmother's bedside as I type these words. I brought the boys with me, knowing that this will almost certainly be the last time they'll see her. My older son is downstairs, ostensibly practicing for his bar mitzvah, although I don't hear him singing. My younger son is sitting cross-legged by my feet, rotating his impossibly loose tooth. It has been "wanting to come out" for days, which looks every bit as much like "wanting to stay in." The room is so quiet we can hear the tooth's root as he turns it. It sounds like a tissue-paper flower. I find it impossible not to imagine my grandmother underground, holding the tissue-paper flower's root as my son innocently wiggles the bulb.

•

It is now two months later. I e-mailed my father and asked what kind of tree it is that my grandmother saw out her window. He responded, "You are probably thinking of the wonderful Japanese maple. Unfortunately it didn't survive. Its replacement is a sycamore, I think. It is still small." In the first suicide note, the author observes, "Yet life is a transitory state, and even trees fall." My son's adult tooth has crowned at the root.

•

When Stephen Hawking presided over the signing of the University of Cambridge's "Declaration on Consciousness," he advanced the idea that, like humans, the animals we eat experience consciousness, "along with the capacity to exhibit intentional behaviors." Generally speaking, we treat other humans humanely because we value their consciousness. It is also the reason many vegetarians don't eat animals. And it is the case we would likely make to a more powerful alien for our own humane treatment.

But what if the alien didn't regard consciousness as sufficient? What if it wanted to know what was *done* with that consciousness? Humans may have "the capacity to exhibit intentional behaviors," but how do we exercise it? It's best not to inflict unnecessary pain on something that can feel pain, but is that an argument for its survival? I find the case against factory farming quite easy to make, whereas the case against meat, per se, has always challenged me.

What is the argument for our survival?

•

My younger son usually asks me to stay by his bedside until I can hear him sleeping. Sometimes, while sitting with him, I think about the last night my mother had with her father, when he kept kissing her and telling her he loved her. Sometimes I already miss what I haven't yet lost, as if I'm staring through the masterpiece at the empty wall behind it.

I often write while waiting for my son's heavy breathing—the sound of my typing assures him that I'm still there—and

I am sitting on the floor of his bedroom now. In increments too small to be detected, he is outgrowing his pajamas, and outgrowing me. I know what I refuse to believe: no empire is big enough, or small enough, to last.

Every time we say "crisis," we are also saying "decision." We must decide what will grow in our place—we must plant our compensation or revenge. Our decisions will determine not only how future generations will evaluate us but whether they will exist to evaluate us at all.

We view the actions of civilians during World War II from the vantage of having won the war. Winning required the ravaging of lives, landscapes, and cultures. Perhaps we look back at those blacked-out houses with admiration, but more likely, we look back and think, *It was the least they could do.*

What if those who came before us had refused to make home-front efforts, and we had lost the war? What if the costs were not extreme, but total? Not eighty million, but two hundred million or more? Not the occupation of Europe, but the domination of the world? Not a Holocaust, but an extinction? If we existed at all, we would look back at a collective unwillingness to sacrifice as an atrocity commensurate with the war itself.

Human populations have driven other human populations to the brink of eradication numerous times throughout history. Now the entire species threatens itself with mass suicide. Not because anyone is forcing us to. Not because we don't know better. And not because we don't have alternatives.

We are killing ourselves because choosing death is more convenient than choosing life. Because the people committing suicide are not the first to die from it. Because we believe that someday, somewhere, some genius is bound to invent a miracle technology that will change our world so that we don't have to change our lives. Because short-term pleasure is more seductive than long-term survival. Because no one wants to exercise their capacity for intentional behavior until someone else does. Until the neighborhood does. Until the energy and car companies do. Until the federal government does. Until China, Australia, India, Brazil, the U.K.—until the whole world does. Because we are oblivious to the death that we pass every day. "We have to do something," we tell one another, as though reciting the line were enough. "We have to do something," we tell ourselves, and then wait for instructions that are not on the way. We know that we are choosing our own end; we just can't believe it.

With every inhale, we take in molecules from Caesar's last gasp. And that of Nina Simone and John Wilkes Booth; Hannah Arendt and Henry Ford; Muhammad, Jesus, Buddha, and Confucius; Roosevelt, Churchill, Stalin, and Hitler; Enrico Fermi, Jeffrey Dahmer, Leonardo da Vinci, Emily Dickinson, Thelonious Monk, Cleopatra, Copernicus, Sojourner Truth, Thomas Edison, and J. Robert Oppenheimer: all the heroes and villains, creators and destroyers.

But most of those exhaled molecules come from ordinary citizens, people like us. I just inhaled my great-aunt saying, "You are so lucky to be leaving." And the silence my grandmother shared with her mother before leaving. And my grandfather telling my mother he loved her in Yiddish that last night. And Frankfurter saying, "My mind, my heart, they are made in such a way that I cannot accept it." More than one hundred billion humans lived on our planet before we did. With each of our inhales, we might ask ourselves if we are worthy of what has been given to us.

We will rise to meet the planetary crisis, or we won't. We will be a wave, or we will drown. If we don't overcome our agnosticism and alter our behavior in the ways that we *know* are necessary, how will our descendants judge us? Will they know that they inherited a battlefield because we were unwilling to turn off our lights?

When my grandmother fled the Nazis as a teenager, she was saving more than just herself: she was saving my mother, my brothers, me, my children, my nieces and nephews, and all the people who will come after us. Life is not always indispensable in the abstract, but it is always indispensable in the particular.

Some humans will survive climate change in scattered, vulnerable populations. But as every other mass extinction on the geological record shows, species that survive one extinction will almost certainly be killed by the next. Their populations and resources dwindle too low for a second resilience.

Even if humans survive global warming, the next proverbial flood will almost certainly end our short reign on this planet. It could be a lethal virus, a drought, an ice age, a volcanic eruption. Perhaps resource scarcity will spark a final war.

At some point, if not on our first try, we will get death right.

And then our planet will orbit unintelligently for the rest of time, an unintelligent rock among unintelligent rocks in an unintelligent universe. The brief experiment with human consciousness—with learning words, planting seeds, sizing the space between monkey bars, twisting loose teeth, trick-or-treating with pillowcases, sliding pencils under casts, making stovepipe hats and beards from construction paper, folding cranes, planting flags, folding poker hands, sharing selfies, wrestling down jealousy, raising pylons to bring electricity to remote communities, raising pylons for beautiful bridges, rowing sailboats in the doldrums, lowering flags to half-mast, struggling to refold maps, sizing engagement rings, launching telescopes to see ever farther into the past, cutting umbilical cords, amortizing, testing the heat of milk against wrists, re-shingling roofs, making way for ambulances, racing hurricanes, preparing wills, misremembering childhood homes, choosing cancer treatments, crumpling inadequate eulogies, setting clocks a few minutes fast to fool oneself, turning off lights to save electricity or a free way of life—will be forever unremembered.

Or perhaps there will be life after us. And perhaps the next inhabitants of what once was our home will arrive soon enough after our disappearance to find artifacts of our tenure: fragments of stone constructions, pieces of plastic, unusual distributions of silicon. Perhaps they will find human handprints left in a cave in southern Argentina, which date to 7300 B.C.E., and human footprints on the moon, and assume that these were equally primitive expressions, or equally sophisticated. Perhaps they will arrange our remains in a museum, accompanied by texts hypothesizing our intentions and what it was like to be human:

> They preferred groups, as small as two. They consumed food when not hungry, engaged in non-procreative sexual activity, and acquired superfluous possessions and knowledge. They struggled with hydration and gravity. They recorded experience with writing implements that disappeared with use. Their hair usually changed color, but their eyes usually did not. They brought their hands together to express approval, and even nonbelievers concealed their feet. They lifted heavy objects, rearranged their teeth. The living needed distance from the dead, but the dead needed proximity to one another. They had names, although very few had unique names.

They had numerous languages and systems of measurement, but no universal language or system of measurement. They paid strangers to touch their backs. They were drawn to chairs, helpless things, privacy and exposure (but nothing in between), reflective minerals, rectangular pieces of glass, organized violence. Each group selected members to worship. They struggled to remain conscious in the dark. They had no armor plating. They sought mirrors to confirm the existence of what they didn't want to see. They had severely limited vision. They passed their death date every year without acknowledging it, and pushed their breath into rubber bladders to commemorate being born. Their needs were too great. Doing nothing to save their kind required the participation of everyone. Every one of them began as a baby, and collectively they were—relative to the history of this planet—extraordinarily young.

Life Note

Dear Boys,

Because I was spending so much time with Bubbe over the past couple of months, she was often on my mind as I wrote this book. It made a certain amount of sense, given the themes: survival, generational responsibility, ends and beginnings. But I also stopped caring if it made sense. There is a refrain in the first suicide note: "To whom do I speak today?" That question is woven through the author's dispute with himself, as if its answer might resolve the dispute. This is no kind of suicide note—it is the opposite of a suicide note—but as I have written, I have returned to the same refrain: *To whom do I speak today?* I started this book with the desire to convince strangers to do something. And while I continue to hope it will do that, as I reach the end, I find myself wanting to address only you.

I was going to take a train to D.C. to see Bubbe this morning, but decided to wait until the weekend, so I could bring you down with me. Grandma called not long after I got back from taking you to school and told me that Bubbe had just passed away. I went straight to Penn Station, slept through the Amtrak ride, and was at Grandma and Pops's house by lunch.

I'm now in Bubbe's room. The funeral home isn't going to come for her body for a couple of hours. I'm sitting beside her bed. Julian and Jeremy were here for a while. Judy, too. Grandma and Pops have been in and out. But now it's just me.

It's the strangest thing not to see the sheet that covers her rising and falling. I keep looking for it, waiting for it, and it keeps not happening. And yet the room still feels as full of her life as ever. It doesn't have to be her heart beating her heartbeat.

•

Your great-grandfather, Bubbe's husband, killed himself a few years after immigrating to America from Europe. I'm not sure if you knew that—or if you knew that you knew it. It's one of those things that is always never talked about. He survived the Holocaust, but he couldn't survive his survival. He died twenty-three years before I was born, and until recently, all that I knew about him was gleaned from a few stories Grandma told me—most of them having to do with how clever and resourceful he was. I didn't know that he killed himself until I was in my thirties. I had to figure it out for myself. In the last few years, Grandma has been much more open about it. Recently, she shared some scraps of paper that were in his pocket when he died—the pieces of a suicide note. The first begins: "My Etele is the best wife in the world."

Isn't it strange how the beginning of his suicide note could also be the beginning of a Valentine's card? The writer Albert Camus once wrote, "What is called a reason for living is also an excellent reason for dying." Your great-grandfather loved his family very much. Sadness and joy aren't opposites of each other. They are each the opposite of indifference.

Maybe one day I'll share with you the notes that Grandma shared with me. They weren't brought together into one text, weren't addressed to anyone, weren't an explanation. I called them a suicide note, but really, what do you call a note like that?

•

Fifteen years after your great-grandfather killed himself, Neil Armstrong landed on the moon. Grandma watched it on television with Bubbe. Don't you wish you had been alive to see that happen? Do you ever think about all the things in the past that you weren't alive to see, or all the things in the future that you won't be alive to see? I just imagined you reading these words when I'm no longer alive.

While Armstrong prepared for his mission, President Nixon's speechwriter prepared some remarks in case the astronauts were stranded on the moon. Here's how that speech, titled "In Event of Moon Disaster," begins:

> Fate has ordained that the men who went to the moon to explore in peace will stay on the moon to rest in peace. These brave men, Neil Armstrong and Edwin Aldrin, know that there is no hope for their recovery. But they also know that there is hope for mankind in their sacrifice. These two men are laying down their lives in mankind's most noble goal: the search for truth and understanding.

If you think about it, how different is it to be an astronaut stranded on the moon than a person living on Earth? You could say that both are stranded. And neither has any "hope for recovery," in the sense that everyone who lives has to die. You could even say that there is "hope for mankind in their sacrifice," if you believe that most people spend their lives contributing to the creation of the world, rather than its destruction. The difference between these two conditions is that, between now and our deaths, only those of us lucky to be stranded on Earth can make ourselves at home.

When Grandma and Bubbe watched the moon landing, they heard Armstrong say what is probably the most famous sentence in human history: "One small step for man, one giant leap for mankind." He had meant to say, "One small step for *a* man," but in that overwhelming moment, the single letter was forgotten. The lowercase *a* is among the physically smallest letters in the alphabet. It's the only lowercase letter that can stand alone, a word unto itself. Maybe he subconsciously omitted it from his statement because he knew that he did not stand alone, that he was not a word unto himself. Probably not.

He'd meant to refer to the single step of an individual, but without the *a*, it was a small step for the human race: "One small step for mankind, one giant leap for mankind."

In order to contribute to the creation of the world, rather than its destruction, an individual must act on behalf of the collective. Humankind takes leaps when individuals take steps.

"In Event of Moon Disaster" was on display at the New York Public Library during the period when I would go there every day to work on my first novel. I looked at it during breaks, aware that it was revealing something to me, but unsure of what.

Five years later, I was about to become a father. I went into a bodega and saw a quart of milk with an expiration date after Sasha's due date, and *believed*, for the first time, that he was going to be born. Even though I'd seen sonograms, felt him moving in Mom's belly, followed the progress of his growth, the birth of a child was too unprecedented, too large, to conceptualize. But I'd many times experienced what happens when milk's expiration passes.

The familiar was my bridge to the unfamiliar, just as the unfamiliar (the improbable terror of being stranded on the moon) was my bridge to the familiar (the improbable fortune of being home on Earth). Nixon's undelivered speech also enhanced my appreciation for what *did* happen—it suddenly seemed miraculous to me that we put people on the moon *and* we brought them home. I had been told the story so many times, an alternative did not occur to me until it was also told. That's why I kept returning to the speech during my breaks—it was addressed to people in another time, to help them contemplate what didn't happen, but it was also addressed to us, to help us contemplate what did.

If we could read an "In Event of Catastrophic Climate Change" speech now, or dig up testimonies from generations to come, or take a meeting with a Karski-equivalent to hear news of unprecedented environmental horror, or pluck a bottle from the ocean bearing a message from our great-great-great-grandchildren, or find scraps of our own suicide notes in the pockets of our clothing, would this evidence bridge the unfamiliar with the familiar and help us understand it? Would we believe what we understood?

•

When I was your age, I used to rummage through Grandma and Pops's closet, hoping to find something I didn't want to find: condoms, marijuana, even a porno. Your grandparents were either more straitlaced than I gave them credit for, or better hiders. The only unexpected thing I ever found was an envelope in Pops's dresser, tucked in the back of a drawer with black socks and squash balls. Across its front he had written: *For my family.*

I didn't dare open it, as that would have given away my pastime, but I didn't feel a need to open it. It's still there. I check on it every now and then. (I actually just checked on it a couple of minutes ago.) I know that he has edited it, because the *For my family* script changes—the size, the color of the pen. While I can't rule out the possibility that the envelope is filled with condoms, marijuana, and porn, or a message that says, "Stop looking through my things!" I've always felt sure of what it contains: a few concise sentences about how much he loved his family, followed by scrupulously organized information about estate planners, insurance policies, bank accounts, safety deposit boxes, cemetery plots, organ donorship, and so on. That's who Pops is. There were years of my life when that drove me crazy. Why couldn't he be more emotional, more expressive? Where was the wildness required by a finite life?

But then I became an adult, and I had you, and now I understand him differently. Pops would seek the advice of an accountant only if he feared he didn't pay enough taxes. He ate red meat twice a day for most of his life but effectively became vegetarian after his parents died of heart attacks. Your grandfather has probably written one hundred unpublished letters to the editor.

To whom are those unpublished letters to the editor speaking?

What do you call a note like the one in his dresser?

•

Forty-three years after Neil Armstrong landed on the moon and said, "One small step for man," an artist named Trevor Paglen launched one hundred photographs into space. They were micro-etched onto an "ultra-archival disc" and encased in a gold-plated shell. His goal was to create images that will last "as long as the sun, if not longer." In 2012, this disc was sent into what is called "stable orbit," meaning at that height—22,236 miles—the effects of gravity and centripetal forces balance out, and, as long as future humans and aliens don't interfere, it will continue to circle Earth until there is no Earth to circle.

Paglen chose photographs that range from photojournalism to near-abstraction, from didactic to impressionistic: the construction of an atomic bomb, orphans seeing the sea for the first time, the sky through blossoming branches, smiling children in a World War II Japanese internment camp, a rocket launch, a stone tablet containing early mathematics.

I don't know what he intended to say with his choices. There is a photo of Trotsky's brain, the set of *Conquest of the Planet of the Apes*, the interior of a factory farm, Ai Weiwei giving the finger to the Eiffel Tower, dinosaur footprints, deep space as seen through the Hubble telescope, the construction of the Hoover Dam, a dandelion, Tokyo at night from above. Paglen's research assistant Katie Detwiler said that they explicitly did not want the project to "be, or appear to be, any attempt to represent humanity—as if that's some stable and monolithic entity." The photographs don't seem to be trying to communicate with another intelligent life-form. Unlike with Carl Sagan's "Golden Record"—which included spoken greetings in fifty-five ancient and modern languages and music from various cultures, as well as images of mathematical and physical equations, the solar system and its planets, DNA, and human anatomy—there seems to be no effort to explain Earth and its inhabitants. The art curator João Ribas called it a "cosmic message in a bottle."

In 1493, on the way back to Spain from the New World, Christopher Columbus's ship was caught in a terrible North Atlantic storm, and he feared that he and his crew would drown. So he wrote a message to King Ferdinand and Queen Isabella describing his discoveries, wrapped the note in a waxed cloth, and put that in a cask, which he then threw into the sea. It's possible that the cask is still floating in the ocean somewhere, like Neil Armstrong's *a* in space, coexisting with all the other things that could have happened but didn't, and all the things that will happen but haven't happened yet: my grandfather's exhalation as he blew out the candles of his forty-fifth birthday cake, the gasps of passengers looking back at what used to be their home, the first breath of the last human.

The one hundred photographs orbiting Earth remind me of the thirty-five thousand papers that the Jews of Warsaw buried during the Holocaust, and of the seeds protected in vaults in case of an agricultural catastrophe. But more than anything, they remind me of the notes in your great-grandfather's pocket: fragments declaring nothing, explaining nothing, questioning nothing. Only arguing.

•

There are many reasons why I will never be an astronaut: a lack of physical fitness, a lack of mental fitness, scientific ignorance. Topping the list is my fear of flying. It's entirely manageable but present every time I get on a plane. These days, it only manifests as concealable panic during turbulence and a runway ritual: as the plane barrels toward takeoff, I say to myself, over and over, "More life . . . More life . . . More life . . ."

To whom am I saying, "More life"? I suppose some part of me believes that if there were a God, and if God could hear me, and could be persuaded to care about me, the simple statement of appreciation of life, and request for more, might be enough to earn me a safe flight. But I don't believe in God. Or at least not in a God who listens, much less responds, to prayer.

I don't believe that the pilot is affected by my prayer. I don't believe that the plane is affected by my prayer. I don't believe that the weather is affected by my prayer.

As I barrel down the runway, repeating "More life . . . More life . . . More life . . . ," I think about my life. I think about it in a way that I don't in any other context. Those thoughts take the form of images. They are not etched onto silicon and sent into stable orbit, where they will exist for hundreds of millions of years. They bloom and wilt in my mind.

I am affected by my prayer.

What do you call a prayer like that? The opposite of a suicide note?

In Flannery O'Connor's story "A Good Man Is Hard to Find," a character is summarized in the following line: "She would have been a good woman, if it had been somebody there to shoot her every minute of her life." If I could spend my entire life barreling down one long runway, I would appreciate what I have so much more than I do now. But if I had to spend my entire life barreling down a runway, I would never have what I appreciate, because I would never be home.

•

I'm back in Bubbe's room, although she's no longer here. Two men from the funeral parlor came around an hour ago to take her away. If being with her body was peaceful, watching her get wrapped up, carried down the stairs and out the front door, was pretty horrible. The thieves who stole the *Mona Lisa* took the painting out the museum's front door, which made the event all the more shocking. How could it have been allowed to happen?

There is a Jewish mourning tradition called *kriah*, which means "ripping." The close relatives of someone who has died rip a piece of their clothing, to symbolize the grief that they feel.

When Bubbe was taken from this room, I felt a ripping—I felt that she was being cut off from me.

Julian and Jeremy are downstairs. Grandma and Pops. Frank's down there, too. I'll join them in a minute, but I wanted a bit more time in here. Bubbe didn't leave this room for the last months of her life. I have become so used to thinking of it as her final room that I all but forgot that it was my childhood bedroom. It was in here that I read *The Catcher in the Rye*, learned my haftorah, heard *OK Computer* for the first time, scrutinized my first pimples, shaved for the first time, read *Lolita*, studied for the SATs, rehearsed one thousand times how I would ask someone to prom. I have forgotten every word of my haftorah, along with the plot of *The Catcher in the Rye*, and I haven't spoken to my prom date in a quarter of a century. But those experiences couldn't have mattered more to me while I was having them, and I'm still inhaling molecules that I exhaled then. We are connected to ourselves and others across space and time, and so we have obligations to ourselves and others, no matter the distances.

What was that artist saying with those images he launched into orbit? We were here? We mattered?

Nobody would question whether Bubbe's life mattered. She was blessed with many things, but she was cursed by history—cursed by her courage and wisdom and resilience—to be larger than life. When she would share her superheroic life story with my Hebrew school class, or even when she spoke in the intimacy of her living room, she was never just my grandma speaking—she was a *representative*, an idea as much as a person. We hugged her because we loved her, but also because we felt, even as children, an obligation to all that our arms could not wrap around.

When Bubbe made sacrifices, the need couldn't have been

more obvious. She walked more than twenty-five hundred miles, endured freezing temperatures, illness, and malnutrition, so that the Nazis wouldn't kill her. And when Grandma and Julian were born, it was obvious why she would clip coupons, and organize pennies into paper rolls, and patch the ripped patches on their well-worn clothing. She needed to keep her children housed and healthy.

Facing climate change requires an entirely different kind of heroism, which is far less intimidating than escaping a genocidal army, or not knowing where your children's next meal will come from, but is perhaps every bit as difficult because the need for sacrifice is *un*obvious.

I grew up in this room, and Bubbe died in this room. This room held some of the most important of our family's dramas. It was our home. But it wasn't built for us. People lived here before us, and there will be others after us. We have obligations to those people—even to people who don't yet exist—just as my brothers and I felt an obligation to the things Bubbe did before we were born, and just as she felt an obligation to us before we existed.

An image just entered my mind, as if I were about to take off in an airplane rather than descend the stairs and rejoin the others. An image as fleeting and enduring as a breath. I'm thinking about when we took a narrow boat down the Erie Canal. You guys were nine and six. Before they gave us the key, we had to attend a twenty-minute orientation. Remember when the instructor asked if we knew how to tie boating knots? And without waiting for our answer, he said, "Well, if you can't tie knots, tie lots?" I loved that. We loved that. We loved scrutinizing the spiral-bound book of nautical maps (despite the canal offering no navigational options), and we loved how fast our ark felt compared with how slow it was—remember all the joggers who passed us on the banks? We loved radioing ahead to the lock masters, making s'mores on the single burner, watching the Monopoly money get taken by a breeze so strong we never saw it land, peeing off the back of the boat simply because we could, revving the meek engine simply because we could, eating the hot-chocolate powder simply because we could, tying lots of knots in the hot rain. You begged to jump into the water from atop the boat. I had to fight against my reflex to protect you from what was perfectly safe. I remember the two of you in the air: Cy's smile, his hands clasped in front of him as if to hold the moment like a firefly. And Sasha's hair, his ribs, his right fist raised in . . . what? In *what*? Triumph over fear? An inherited fight-or-flight reflex that predates *Homo sapiens*? Love of life?

"To whom do I speak today?" the author of the first suicide note repeats over and over as he enumerates the arguments for giving up. The soul instructs him to "cling to life," comparing death with "taking a man out of his house."

It is not enough to say that we want more life; we must refuse to stop saying it. Suicide notes are written once; life notes must always be written—by having honest conversations, bridging the familiar with the unfamiliar, planting messages for the future, digging up messages from the past, digging up messages from the future, disputing with our souls and refusing to stop. And we must do this together: everyone's hand wrapped around the same pen, every breath of everyone exhaling the shared prayer. "Thus we shall make a home together," the soul concludes at the end of the suicide note, perhaps beginning its opposite. Each of us arguing with ourselves, we shall make a home together.

Appendix: 14.5 percent / 51 percent

Two of the most frequently cited reports on animal agriculture's contributions to the environmental crisis—*Livestock's Long Shadow* from the Food and Agriculture Organization (FAO) of the United Nations in 2006 and "Livestock and Climate Change" from the Worldwatch Institute in 2009—provide two different sets of numbers on what is one of the most important data points in all environmental science: the percentage of greenhouse gas emissions produced by the livestock sector. This number is a species of super-statistic that incorporates and simplifies a vast complexity, and is the most straightforward argument for why changing our relationship with animal products is so crucial.

Livestock's Long Shadow was the first report of its kind to gain widespread attention, and when it claimed that animal agriculture caused 18 percent of global greenhouse gas emissions, it attracted applause and criticism. Mostly, however, it triggered alarm: 18 percent was more than the entire transport sector combined. It was therefore surprising when, in 2009, the Worldwatch Institute published its report in response to *Livestock's Long Shadow*, claiming that animal agriculture was not responsible for 18 percent of annual greenhouse gas emissions worldwide, but in fact *at least* 51 percent. "If this argument is right," state the authors in their introduction, "it implies that replacing livestock with better alternatives would be the best strategy for reversing climate change." They recommend a 25 percent reduction in the number of livestock worldwide, which, they clarify, "can be made to happen in select locations so that livestock populations in poor rural communities would remain entirely intact."

It's worth pausing on how those two very different numbers were reached, as it is not only of great scientific importance but also reveals how our understanding of our planet can be so out of whack with reality.

Robert Goodland and Jeff Anhang authored the Worldwatch report, which is called "Livestock and Climate Change: What If the Key Actors in Climate Change

Are . . . Cows, Pigs, and Chickens?" (Their ellipsis, not mine.) Those eager to challenge the credibility of the study, including the authors of *Livestock's Long Shadow*, claim that it was not peer-reviewed. Others—including Anhang and Goodland themselves—maintain that it was, and that the self-published FAO report was not.

Jeff Anhang works for the World Bank Group's International Finance Corporation. Robert Goodland, who passed away in 2014, was an ecologist, a professor, and a lead environmental adviser to the World Bank Group. He had a Ph.D. in environmental science and served as president of the International Association of Impact Assessment. After he retired from the Worldwatch Institute in 2001, he directed environmental and social impact assessment studies on more than a dozen worldwide projects. In other words, he was not an animal rights guy and not a hobbyist.

In a 2012 piece he wrote for *The New York Times*, Goodland said the following:

> The key difference between the 18 percent and the 51 percent figures is that the latter accounts for how exponential growth in livestock production (now more than 60 billion land animals per year), accompanied by large scale deforestation and forest-burning, have caused a dramatic decline in the Earth's photosynthetic capacity, along with large and accelerating increases in volatilization of soil carbon.

In the executive summary of their Worldwatch report, Goodland and Anhang expand this point, arguing that the forgone carbon absorption associated with the livestock industry's deforestation should be accounted for:

> The FAO counts emissions attributable to changes in land use due to the introduction of livestock, but only the relatively small amount of GHGs [greenhouse gases] from changes each year. Strangely, it does not count the much larger amount of annual GHG reduction from photosynthesis that are forgone by using 26 percent of land worldwide for grazing livestock and 33 percent of arable land for growing feed, rather than allowing it to regenerate forest. By itself, leaving a significant amount of tropical land used for grazing livestock and growing feed to regenerate as forest could potentially mitigate *as much as half (or even more) of all anthropogenic GHGs*.

In their 2010 follow-up, which they published to answer questions from the public, Goodland and Anhang defend the choice to include forgone carbon absorption, stating: "We believe that counting a forgone reduction of any magnitude is valid because it has exactly the same effect as an increase in emissions of the same magnitude."

Goodland and Anhang identify and correct many other uncounted, overlooked, and misallocated livestock-related greenhouse gas emissions in the FAO report. Among them: overlooked land use, undercounted methane, and undercounted livestock. They also claim that the FAO overapplied data from Minnesota, which is a problem because livestock operations there are more efficient than they are in the developing world, where the sector is expanding the fastest. Goodland and Anhang write that in some sections, *Livestock's Long Shadow* "uses lower numbers than appear in FAO statistics and elsewhere." In addition, the FAO fails to account for deforestation in some countries (like Argentina) and omits farmed fish from its calculation.

Finally, the FAO does not account for the "substantially higher amount of GHGs attributable" to livestock products versus plant-based alternatives. Livestock products need to be refrigerated, which requires fluorocarbons—compounds that have a global warming potential (GWP) up to several thousand times higher than that of CO_2. Cooking animal products is more GHG-intensive than cooking alternative foods is. The FAO overlooks emissions associated with liquid waste disposal and animal by-products—like bones, fur, fat, and feathers—which are either disposed of or distributed.

Goodland and Anhang also point out that the FAO authors used outdated information. For example, the FAO report used a methane GWP of 23 on a one-hundred-year timeframe, when the Intergovernmental Panel on Climate Change supports a GWP of 25. (On a twenty-year time frame, methane's GWP is 72.) Using the outdated number, the FAO calculates that methane is responsible for 3.7 percent of worldwide GHG emissions. "When methane GWP is adjusted to the 20-year timeframe," Goodland and Anhang counter, "livestock methane is responsible for 11.6 percent of worldwide GHGs. Recalculation increases GHGs from livestock by 5,047 million tons of CO_2e." Put most simply, when you adjust for methane's extra heat-trapping power over a shorter time frame, it contributes a higher proportion of emissions.

Imagine you were outside on a hot summer day and someone handed you a blanket. He tells you that you have to wear this blanket for ten hours. For the first two hours, it will be an electric blanket, three times more powerful than a regular blanket. Then the electricity will turn off. Asking whether to calculate emissions based on a twenty-year time frame or a hundred-year time frame is the difference between asking, "How hot did this blanket make you for the first two hours?" and "How hot did this blanket make you overall?"

Now imagine that in theory your body could handle the overall heat, but the overall heat was irrelevant, because those first two hours were so hot you experienced heatstroke and had to go to the hospital. Because we have less than twenty years to address climate change, some scientists argue that we should calculate the short-term GWPs of greenhouse gases. Two degrees of global warming is a "tipping point," after which positive feedback loops might trigger uncontrollable warming, effectively killing us.

The FAO also "uses citations for various aspects of GHGs attributable to livestock dating back to such years as 1964, 1982, 1993, 1999, and 2000. Emissions today would be much higher."

Another major source of carbon dioxide emissions was not counted in *Livestock's Long Shadow*: livestock respiration. The FAO authors Henning Steinfeld and Tom Wassenaar claim that livestock respiration should *not* be counted, arguing: "Emissions from livestock respiration are part of a rapidly cycling biological system, where the plant matter consumed was itself created through the conversion of atmospheric CO_2 into organic compounds." Since the emitted and absorbed quantities are considered to be equivalent, livestock respiration is not considered to be a net source under the Kyoto Protocol. (The Kyoto Protocol set internationally binding emission reduction targets. It was adopted in 1997, and its first commitment period began in 2008.)

Goodland and Anhang, however, put forth a highly persuasive argument in favor of counting livestock respiration and claim that it is considered a net source under the Kyoto Protocol. They point out that livestock are not essential for human life and that huge swaths of the human population eat few to no animal products. "Today," Goodland and Anhang state, "tens of billions more livestock are exhaling CO_2 than in pre-industrial days, while Earth's photosynthetic capacity . . . has declined sharply as forest has been cleared." They then cite an estimate that CO_2 from livestock respiration accounts for 21 percent of anthropogenic GHGs worldwide. In the follow-up piece, Goodland and Anhang add that "carbon flowing into the atmosphere from animal respiration and soil oxidation exceeds that absorbed due to photosynthesis by 1–2 billion tons per year."

In short, unlike wild buffalo roaming precolonial America, industrial livestock operations are *not* a part of a natural carbon cycle—especially considering how many of the planet's carbon-absorbing forests have been destroyed either to make room for the animals or to make room for growing the corn and soy to feed them—and it is no longer possible for livestock to live in natural harmony with the planet's photosynthetic processes.

Which is why, in addition to accounting for cow gas, Goodland and Anhang argue that it is also necessary to account for forgone carbon absorption caused by livestock-motivated deforestation. This is an especially relevant toll because the livestock industry is clearing the kinds of forests that have the greatest photosynthetic capacity:

> Growth in markets for livestock products is greatest in developing countries, where rainforest normally stores at least 200 tons of carbon per hectare. Where forest is replaced by moderately degraded grassland, the

tonnage of carbon stored per hectare is reduced to 8. On average, each hectare of grazing land supports no more than one head of cattle, whose carbon content is a fraction of a ton. In comparison, over 200 tons of carbon per hectare may be released within a short time after forest and other vegetation are cut, burned, or chewed.

Using their new calculations, the Worldwatch authors claim that animal agriculture is in fact responsible for *at least* 32,564 million tons of annual GHG emissions in CO_2e, compared with the 7,516 million tons estimated by the FAO.

In 2011, a scathing commentary on the Worldwatch study ran in *Animal Feed Science and Technology*. In their response, titled "Livestock and Greenhouse Gas Emissions: The Importance of Getting the Numbers Right," the authors (Mario Herrero et al.) repeatedly cite the "widely recognized" and "well documented" *Livestock's Long Shadow* and then attack and dismantle the credibility of the Worldwatch report, claiming it was not peer-reviewed and implying that it includes "major deviations from international protocols." The commentary fails to acknowledge that two of its authors were also authors of *Livestock's Long Shadow*.

When contacted by *The Philadelphia Inquirer* in 2012, Anhang claimed that their study was, in fact, peer-reviewed. "As an employee of the World Bank Group's International Finance Corporation, Jeff Anhang was required to have peer review on any report he had his name on," writes the journalist Vance Lehmkuhl, who goes on to say that he "pressed Goodland and Anhang on this question and received details on the researchers and institutions who had reviewed the Worldwatch draft prior to publication, as well as those that have cited it subsequently. On the other hand, Livestock's Long Shadow may or may not have been peer-reviewed. FAO cites no such process (nor does the Herrero et al. commentary) and I was unable to find any reference to peer review in any coverage of LLS. I emailed Mario Herrero for clarification on this but have received no reply."

Anhang pointed out to me that the Worldwatch work was actually peer-reviewed twice—once before appearing as an article published in *Animal Feed Science and Technology*, and once again in the form of post-publication peer review in a 2010 Worldwatch article.

Later in 2011, Goodland and Anhang wrote a *response to the response to their report* (which itself was a response to another report), in which they counterargue the counterarguments to their original counterargument.

Then, in 2012, Goodland published a piece in *The New York Times*, "FAO Yields to Meat Industry Pressure on Climate Change." "Frank Mitloehner," Goodland writes, "known for his claim that 18 percent is much too high a figure to use in the U.S., was announced last week as the chair of a new partnership between the meat industry and FAO. FAO's new partners include the International Meat Secretariat and International Dairy Federation. Their stated objective is to 'assess the environmental performance of the livestock sector' and 'to improve that performance.'" Goodland claims that this new partnership isn't surprising, considering the lead author and coauthor of *Livestock's Long Shadow* "later wrote to prescribe more factory farming, not less, and no limit on meat," whereas the World Bank urges institutions to "'avoid funding large-scale commercial, grain-fed feedlot systems and industrial milk, pork, and poultry production.'"

Sure enough, in 2013, the FAO released a new report, stating: "With emissions estimated at 7.1 gigatonnes CO_2-eq per annum, representing 14.5 percent of human-induced GHG emissions, the livestock sector plays an important role in climate change."

•

So, 14.5 percent or 51 percent? I don't think either number is accurate, but I find the higher end far more persuasive. And I'm not alone. A 2014 UN General Assembly report elevated the 51 percent assessment above the FAO's estimate: "The precise figures remain debated, but there is no doubt in the scientific community that the impacts of livestock production are massive." UNESCO, another UN agency, also published a report favoring the 51 percent estimate above that of the FAO. The UNESCO authors write that the Worldwatch calculation "represents an enormous shift in perspective, and further strengthens the evidence for the relationship between meat production and effects on climate change."

I had a lengthy e-mail exchange with Jeff Anhang, inviting him to respond to the various critiques of his calculations. Finally, I asked what, in his opinion, we need to do to meet the goals of the Paris accord.

"It seems impossible to reverse climate change by capping fossil fuels," he wrote. "That's because the amount of renewable energy infrastructure needed to stop climate change has been estimated by the International Energy Agency to cost at least $53 trillion and take at least twenty years, by which time it's projected to be too late to reverse climate change. In contrast, replacing animal products with alternatives offers a unique dual opportunity to reduce greenhouse gas emissions quickly while freeing up land to enable more trees to capture excess atmospheric carbon in the near term. So replacing animal products with alternatives seems to be the only pragmatic way to reverse climate change before it is too late."

Notes

I. UNBELIEVABLE

3 *The oldest suicide note was written in ancient Egypt*: Erman, *Ancient Egyptians.*

3 *The first line reads*: "Dialogue of a Man with His Soul."

4 I am not the least afraid: Kearl, *Endings*, 49.

4 *Armstrong would leave that boot on the moon*: Bethge, "Urine Containers, 'Space Boots' and Artifacts Aren't Just Junk."

4 *When Alex, the African grey parrot*: Carey, "Parrot Who Had a Way with Words."

5 *The Roman Empire*: Taagepera, "Size and Duration of Empires."

7 *They weren't, themselves*: Gannon, *Operation Drumbeat.*

8 *The SS* Robert E. Peary: American Merchant Marine at War, "Liberty Ship SS Robert E. Peary."

8 *Lingerie factories began making*: Rosener, *Women in Industry.*

8 *Retirees, women, and students*: Ossian, *Forgotten Generation*, 73.

8 *Celebrities encouraged the purchase*: Fitch, "Julia Child."

8 *Top marginal tax rates*: Roosevelt, "Executive Order 9250."

8 *Gasoline was severely regulated*: Perrone and Handley, "Home Front Friday."

8 *U.S. government posters*: Pursell, "When You Ride ALONE."

9 *Food was rationed*: George C. Marshall Foundation, "National Nutrition Month."

9 *In the U.K., people were eating*: Collingham, *Taste of War.*

9 *This collective act of belt-tightening*: British Nutrition Foundation, "How the War Changed Nutrition."

9 *"Their weapons are the panzer forces"*: Walt Disney Productions, *Food Will Win the War.*

9 *"Not all of us can have"*: Roosevelt, "Fireside Chat 21."

10 *more than four trillion dollars*: Daggett, *Costs of Major U.S. Wars.*
11 *Imagine if the remaining 10.5 million Jews*: Sifferlin, "Global Jewish Population."
12 *The chief threat to human life*: Vidal, "Protect Nature."
13 *We know climate change*: Milman, "Climate Change."
13 *or contributing to a polar vortex*: Rice, "Yes, Chicago Will Be Colder Than Antarctica."
13 *And we find it hard*: Rebuild by Design, "The Big U."
14 *Claudette Colvin*: Rumble, "Claudette Colvin."
15 *Like the iconic photographs*: Poirier, "One of History's Most Romantic Photographs."
15 *photo of Rosa Parks*: Rothman and Aneja, "Rosa Parks."
15 *And as she later acknowledged*: Sullivan, "Bus Ride."
16 *As the marine biologist and filmmaker Randy Olson put it*: Revkin, "Global Warming."
16 *"The climate crisis is also a crisis of culture"*: Ghosh, *The Great Derangement*, 9.
17 *"The Bund leader came up to me"*: Lewin and Bartoszewski, *Righteous Among Nations.*
19 *In a recent study, the UCLA psychologist Hal Hershfield*: Hershfield, "Better Decisions."
20 *It has been widely demonstrated*: Ibid.
20 *Researchers have described a number of "sympathy biases"*: Sudhir et al., "Sympathy Biases."
20 *Combining all these tactics*: Institute for Operations Research and the Management Sciences, "Carefully Chosen Wording."
20 *They feel abstract, distant*: Ballew et al., "Global Warming as a Voting Issue."
20 *As the journalist Oliver Burkeman put it*: Burkeman, "Climate Change Deniers."
20 *So-called climate change deniers*: Nuccitelli, "Climate Consensus"; NASA, "Scientific Consensus."
23 *"What is knowledge?"*: Brody, "The Unicorn and 'The Karski Report.'"
25 *When explaining why he did what he did*: Huicochea, "Man Lifts Car."
25 *Boyle, who was not a weight lifter*: Wise, *Extreme Fear*, 25–27.
26 *In Moscow in the early 2010s*: "Russia's Rich."
27 *In London, when a Piccadilly nightclub*: Rennell, "Blitz 70 Years On."
30 *A study of Germany's Bundesliga soccer league*: Dohmen, "In Support of the Supporters?"
30 *"To mobilize people"*: Marshall, *Don't Even Think About It*, 57.
34 *In 2018, despite knowing more*: "Global Carbon Dioxide Emissions Rose."
36 *A little less than two centuries*: Brandt, "Google Divulges Numbers."
36 *Researchers recently identified*: Shah, "Addicted to Selfies."
36 *They named it "chronic selfitis"*: Lee, "What Is 'Selfitis'?"
37 *Explaining the rise of MSNBC*: Schwartz, "MSNBC's Surging Ratings."

37 *A recent study published in* Environmental Science and Technology: Babaee et al., "Electric Drive Vehicles."

37 *an average person's vehicle emissions*: Schiller, "Buying a Prius."

38 *Do the children getting vaccines*: Leskin, "13 Tech Billionaires."

38 *Do the children dying*: Kotecki, "Jeff Bezos."

39 *One after the other, individual bees*: Kastberger et al., "Social Waves in Giant Honeybees."

39 *But it* is *the case that bee populations*: Xerces Society for Invertebrate Conservation, "Bumblebee Conservation."

40 *From China to Australia to California*: Wilder, "Bees for Hire"; Pensoft Publishers, "Bees, Fruits and Money."

40 *A photographer who documented this process said*: Williams, "Shrinking Bee Populations."

42 *Ninety-six percent of American families*: Harris Poll, "Carrying on Tradition."

42 *That is higher than the percentage*: Delta Dental, *2014 Oral Health and Well-Being Survey*, 8.

42 *have read a book in the last year*: Perrin, "Who Doesn't Read Books in America?"

42 *or have ever left the state*: Nicholas, "Home State."

42 *If Americans had set a goal*: University of Illinois Extension, "Turkey Facts."

43 *A recent study by the Stockholm School of Economics*: Mellström and Johannesson, "Crowding Out in Blood Donation."

43–44 *But pilgrims who had participated*: Khan et al., "Collective Participation."

44 *"Engaging in longer and more satisfying"*: Muise et al., "Post Sex Affectionate Exchanges."

44 *After the grocery store Pay and Save*: Moss, "Nudged to the Produce Aisle."

44 *In countries where citizens have to opt in*: Davidai et al., "Default Options for Potential Organ Donors."

44 *playful stickers*: Thaler and Sunstein, "Easy Does It."

45 *About 37 percent of registered voters*: United States Elections Project, "2014 November General Election Turnout Rates."

45 *In the 2016 presidential election*: United States Elections Project, "2016 November General Election Turnout Rates."

47 *He never denied*: Gallagher, *FDR's Splendid Deception.*

47–48 *In 1952, after successfully inoculating*: Shampo and Kyle, "Jonas E. Salk."

48 *Salk began human testing*: Salk Institute for Biological Studies, "About Jonas Salk."

48 *The Thanksgiving that we celebrate today*: Lincoln, "Proclamation of Thanksgiving."

50 *Yet the smoking rate in America*: Dennis, "Who Still Smokes in the United States."

50 *As early as 1949*: Moore, "Nine of Ten Americans View Smoking as Harmful."

50 *than there are people in Canada*: Worldometers, "Canada Population (Live),"

37.16 million; accessed February 18, 2019, http://www.worldometers.info/world-population/canada-population/.

50 *Why would someone*: Holford et al., "Tobacco Control."

50 *Despite the rising cost of cigarettes*: Truth Initiative, "Why Are 72% of Smokers from Lower-Income Communities?"

51 *"It was obviously a help in getting teenagers"*: McKie, "A Jab for Elvis Helped Beat Polio."

52 *In a news release*: Scheiber, "Google Workers"; Wakabayashi et al., "Google Walkout."

54 *"Tell your parents not to ruin"*: Guggenheim, *An Inconvenient Truth.*

55 *According to a 2017 analysis*: Wynes and Nicholas, "Climate Mitigation Gap."

56 *"the most important contribution every individual"*: Frischmann, "100 Solutions to Reverse Global Warming."

58 *All these emissions need to fall*: Gates, "Climate Change."

58 *Even if we are miraculously able to achieve it*: Raftery et al., "Less Than 2°C Warming by 2100 Unlikely."

58 *Sea levels will rise*: Worland, "These Cities May Soon Be Uninhabitable."

58 *flooding coastlines*: Schleussner et al., "Differential Climate Impacts."

58 *143 million people are projected to become climate migrants*: Parker, "Climate Migrants."

58 *Armed conflict will increase*: Burke et al., "Climate and Conflict."

58 *Greenland will tip into irreversible melt*: Robinson et al., "Greenland Ice Sheet."

58 *Between 20 and 40 percent of the Amazon*: Intergovernmental Panel on Climate Change, *Global Warming of 1.5°C.*

58 *The European heat wave of 2003*: Di Leberto, "Summer Heat Wave."

59 *Human mortality will dramatically*: Mann and Kump, *Dire Predictions*, 50–162.

59 *The number of people at risk of malaria*: World Health Organization, "Climate Change and Human Health"; World Bank, *Turn Down the Heat.*

59 *Four hundred million people*: Wallace-Wells, *Uninhabitable Earth*, 12.

59 *Warmer oceans will irreparably damage*: Schleussner et al., "Differential Climate Impacts."

59 *Half of all animal species*: Meixler, "Half of All Wildlife."

59 *A total of 60 percent of all plant species*: World Wildlife Fund, "Wildlife in a Warming World."

59 *Wheat yields will be reduced by 12 percent*: Zhao et al., "Temperature Increase Reduces Global Yields."

59 *Global GDP per capita will drop*: Wallace-Wells, *Uninhabitable Earth*, 12.

59 *The few experts*: Raftery et al., "Less Than 2°C Warming by 2100 Unlikely."

61 *The operation is happening now*: Tillman, *D-Day Encyclopedia.*

61 *sewing hundreds of dummies*: Morton, "Object of Intrigue."

69 *Perhaps we could argue*: Scranton, "Raising My Child."

II. HOW TO PREVENT THE GREATEST DYING

75 *The average global temperature*: Jouzel et al., "Antarctic Climate Variability"; Prairie Climate Center, "Four Degrees of Separation"; NASA Earth Observatory, "Today's Warming."

75 *Fifty million years ago*: Eberle et al., "Seasonal Variability in Arctic Temperatures"; Scott and Lindsey, "What's the Hottest Earth's Ever Been?"; Jardine, "Patterns in Palaeontology."

76 *The most lethal mass extinction*: Penn et al., "Temperature-Dependent Hypoxia"; New York University, "Siberian Volcanic Eruptions"; Zimmer, "Sudden Warming."

76 *Many scientists call the geological age*: Welcome to the Anthropocene, www.anthropocene.info.

77 *Taking into account natural mechanisms*: Ritchie, "Exactly How Much Has the Earth Warmed?"; NASA Earth Observatory, "Is Current Warming Natural?"; Union of Concerned Scientists, "How Do We Know That Humans Are the Major Cause of Global Warming?"

78 *Humans represent 0.01 percent*: Carrington, "Humans Just 0.01% of All Life"; Bar-On et al., "Biomass Distribution on Earth."

79 *Globally, humans use 59 percent*: Steinfeld et al., *Livestock's Long Shadow*, xxi.

79 *One-third of all the fresh water*: Gerbens-Leenes et al., "Water Footprint of Poultry, Pork and Beef."

79 *while only about one-thirtieth*: Hoekstra et al., "Water Footprint of Humanity."

79 *Seventy percent of the antibiotics*: Ritchie, "How Do We Reduce Antibiotic Resistance from Livestock?"

79 *There are approximately thirty*: Compassion in World Farming, *Strategic Plan 2013–2017*; Fishcount, "Farmed Fish."

80 *Before the Industrial Revolution*: Zijdeman and Ribeira da Silva, "Life Expectancy at Birth."

80 *It took two hundred thousand years*: United Nations Department of Economic and Social Affairs, Population Division, "World Population Prospects."

80 *Every day, 360,000 people*: Lamble, "How Many People Can the Earth Sustain?"

81 *In 1820, 72 percent*: American Farm Bureau Federation, "Fast Facts About Agriculture; "Farm Population Lowest Since 1850's"; United States Bureau of Labor Statistics, "Employment Projections Program."

81 *Like the video game console*: Brookhaven National Laboratory, "The First Video Game?"

81 *Between 1950 and 1970*: Ganzel, "Shrinking Farm Numbers"; United States Bureau of the Census, "Census of Agriculture, 1969 Volume II."

81 *During that time, the size of the average chicken*: Zuidhof et al., "Commercial Broilers."

81 *In 1966, distorting contact lenses*: Wise and Hall, Distorting contact lenses for animals, U.S. Patent 3,418,978.

81 *The lenses were considered too burdensome*: "Super-Sizing the Chicken."

82 *In 2018, more than 99 percent*: Sentience Institute, "US Factory Farming Estimates."

83 *The current level of meat and dairy consumption*: Steinfeld et al., *Livestock's Long Shadow*; Durisin and Singh, "Americans' Meat Consumption."

83 *Humans eat sixty-five billion*: Gorman, "Age of the Chicken."

83 *On average, Americans consume*: Pasiakos et al., "Animal, Dairy, and Plant Protein Intake."

83 *People who eat diets high in animal protein*: Levine et al., "Low Protein Intake."

83 *Smokers are three times as likely*: Centers for Disease Control and Prevention, "Tobacco-Related Mortality."

83 *In America, one out of every five meals*: Mooallem, "Last Supper."

84 *We are currently in the Quaternary glaciation*: National Centers for Environmental Information, "Glacial-Interglacial Cycles."

84 *According to models of cyclical climate change*: Joint Study for the Atmosphere and the Ocean, "PDO Index"; Physikalisch-Meteorologische Observatorium Davos / World Radiation Center (PMOD/WRC), "Solar Constant"; National Weather Service Climate Prediction Center, "Cold and Warm Episodes by Season."

84 *Nine of the ten warmest years*: National Centers for Environmental Information, "Global Climate Report."

84 *During the Great Dying*: Solly, "The 'Great Dying.'"

84 *Humans are now adding greenhouse gases*: Wallace-Wells, "Uninhabitable Earth."

85 *Life on Earth depends*: Ma, "Greenhouse Gases."

85 *CO_2 accounts for 82 percent*: United States Environmental Protection Agency, *U.S. Greenhouse Gas Emissions and Sinks.*

85 *For the eight hundred thousand years before the Industrial Revolution*: United States Environmental Protection Agency, "Climate Change Indicators."

85 *Since the Industrial Revolution*: American Chemical Society, "Greenhouse Gas Changes."

86 *Animal agriculture is responsible*: Steinfeld et al., *Livestock's Long Shadow.*

87 *One of the most powerful feedback loops*: National Snow and Ice Data Center, "All About Sea Ice: Albedo."

88 *The former United Nations climate chief*: Harvey, "Dangerous Climate Change."

89 *Methane has 34 times*: Intergovernmental Panel on Climate Change, *Climate Change 2013*, chap. 8, 711–14, table 8.7.

90 *Because they are primarily created*: Clark, "Greenhouse Gases."

91 *This is similar to the earth's photosynthetic capacity*: Strain, "Planet's Vegetation."

91 *about one-quarter of anthropogenic emissions:* Climate and Land Use Alliance, "The Earth's Climate."

91 *The more forests we destroy*: Erb et al., "Global Vegetation Biomass."

91 *Allowing tropical land currently used for livestock to revert to forest*: Goodland and Anhang, "'Livestock and Climate Change': Critical Comments and Responses," 13.

92 *Trees are 50 percent carbon*: Food and Agriculture Organization of the United Nations, "Deforestation Causes Global Warming."

92 *Forests contain more carbon*: Climate and Land Use Alliance, "The Earth's Climate."

92 *The cutting and burning of forests*: "Deforestation and Its Extreme Effect on Global Warming."

92 *"By most accounts"*: Ibid.

92 *About 80 percent of deforestation*: Food and Agriculture Organization of the United Nations, "Deforestation Causes Global Warming."

92 *Every year, wildfires in California*: United States Department of the Interior, "2018 California Wildfires."

93 *It has been estimated that Bolsonaro's policy*: Wallace-Wells, "One Man."

93 *Animal agriculture is responsible*: Margulis, "Causes of Deforestation."

94 *As they digest food*: United States Environmental Protection Agency, "Enteric Fermentation."

94 *Livestock are the leading source of methane emissions*: United States Environmental Protection Agency, *U.S. Greenhouse Gas Emissions and Sinks*, 5-1.

94 *Nitrous oxide is emitted by livestock urine*: Steinfeld et al., *Livestock's Long Shadow.*

94 *Livestock are the leading source*: Ibid.

94 *Animal agriculture is the leading cause*: World Wildlife Fund, "Forest Conversion."

94 *According to the United Nations Framework Convention*: Gates, "Climate Change."

95 *The Food and Agriculture Organization*: Steinfeld et al., *Livestock's Long Shadow.*

95–96 *"livestock (like automobiles) are"*: Goodland and Anhang, "Livestock and Climate Change," 12.

97 *Scientists estimate that to keep*: McKibben, "Global Warming's Terrifying New Math."

97 *According to a recent Johns Hopkins University report*: Kim et al., "Mitigating Catastrophic Climate Change."

98 *The most optimistic estimates suggest*: Jacobson and Delucchi, "Path to Sustainable Energy," 64.

98 *"The maths is brutally clear"*: Harvey, "Dangerous Climate Change."

98 *The four highest-impact things*: Wynes and Nicholas, "Climate Mitigation Gap."

99 *Eighty-five percent of Americans drive*: Chase, "Car-Sharing."

99 *Businesses must rely more*: "U.S. Air Passengers' Main Trip Purposes."

100 *Pounds of CO_2e associated with a serving*: Center for Sustainable Systems, "Carbon Footprint Factsheet."

100 *Not eating animal products for breakfast and lunch*: Kim et al., "Country-Specific Dietary Shifts."

101 *To meet the Paris accord's two-degree goal*: Girod et al., "Climate Policy."

101 *While citizens of different countries*: "Carbon Emissions per Person, by Country."
101 *the average Frenchman's*: Ibid.
101 *the average Bangladeshi's*: Ibid.
101 *the average global citizen*: Ibid.
101 *Not eating animal products for breakfast and lunch*: This figure accounts for 235 country-specific practices, including feed composition, feed conversion ratios, and manure management techniques. It accounts for conversion to grassland from forest, but does not account for soil carbon losses from livestock management (desertification). It is backed up by a forthcoming study by Raychel Santo and Brent Kim of Johns Hopkins University.

III. ONLY HOME

106 *If Mars was once inhabited*: Weisman, "Earth Without People."
111 *According to the cognitive psychologist*: Dahl, "Why Can't You Smell Your Own Home?"
113 *You have to achieve at least*: Reinert, "Blue Marble Shot."
113 *"Photo sessions were scheduled events"*: Ibid.
113 *Apollo 17 was the last*: Smithsonian National Air and Space Museum, "Apollo to the Moon."
113 *As the Apollo 8 astronaut William Anders*: New Mexico Museum of Space History, "International Space Hall of Fame."
114 *Many have attributed the rise of the environmental movement*: Kluger, "Earth from Above"; United States Environmental Protection Agency, "Earthrise."
114 *It wasn't when he landed on the moon*: Nardo, *Blue Marble*, 46.
114 *Awe is inspired by two things*: Shaw, "Overview Effect."
114 *The overview effect changes people*: Goldhill, "Astronauts Report an 'Overview Effect.'"
114 *One, Edgar Mitchell*: Institute of Noetic Sciences, "Our Story."
114 *Since Yuri Gagarin*: WorldSpaceFlight, "Astronaut/Cosmonaut Statistics."
115 *According to the space engineer Isaac DeSouza*: Ferreira, "Seeing Earth from Space."
115 *The company's goal*: www.SpaceVR.co.
115 *Commenting on this possibility*: Berger, "Viewing Earth from Space."
115 *Describing his nonvirtual experience*: Garan, *Orbital Perspective*.
116 *The earliest spectacles*: Mortimer, "Mirror Effect."
117 *The rise of the glass mirror*: Ibid.
117 *When babies begin*: Rochat, "Five Levels of Self-Awareness."
117 *A recent addition to this list*: Buehler, "Tiny Fish."
118 *Cleaner wrasse live*: Aton, "Earth Almost Certain to Warm."
118 *George H. W. Bush*: Worland, "Climate Change."
118 *He pledged to*: Thompson, "Timeline."
118 *That year, forty-two senators*: Rich, "Losing Earth."

119 *According to the EPA*: United States Environmental Protection Agency, "International Treaties and Cooperation."

119 *About six years before*: Cushman, "Climate Research Budget."

119 *In his investigative article*: Rich, "Losing Earth."

120 *Bush promised that "{his} Administration's climate change policy"*: Office of the Press Secretary, "President's Statement on Climate Change."

120 *That same year, he established*: United States Climate Change Science Program, "Climate Change Research Initiative."

120 *In his speech discussing why*: Office of the Press Secretary, "President Bush Discusses Global Climate Change."

120 *In America, it is easier than ever*: Greshko et al., "How Trump Is Changing the Environment."

120 *Although his administration*: Lavelle, "Obama's Climate Legacy."

120 *Recently, supposedly progressive hotbeds*: McKibben, "Up Against Big Oil."

121 *Abroad, the French turned out*: Nossiter, "France Suspends Fuel Tax Increase."

121 *Almost fifty years after the Apollo 17 astronauts*: Matthews, "Climate Change Skepticism."

123 *"I am convinced that humans need to leave Earth"*: Knapton, "Human Race Is Doomed."

123 *If the 7.5 billion people on the planet*: McDonald, "How Many Earths Do We Need?"

124 *The GFN estimates*: Coren, "Earth's Natural Resources."

124 *Seventy-three percent of American consumers*: Calfas, "Americans Have So Much Debt."

124 *Humanity has a DTI ratio*: McDonald, "How Many Earths Do We Need?"

124 *In fact, twenty-one youth plaintiffs*: Our Children's Trust, "Juliana v. U.S.-Climate Lawsuit"; Conca, "Children Change the Climate."

125 *Many economists argue*: Allison, "Financial Health of Young America."

127 *We may figure out how*: Dunn, "1,000 Passenger Ships."

127 *minus eighty degrees Fahrenheit*: NASA, "Mars Facts."

128 *Responding to techno-interventions*: National Research Council, *Climate Intervention*, 9.

129 *According to Project Drawdown*: Project Drawdown, "Solutions."

130 *Membership in the NAACP*: Virginia Museum of History and Culture, "Turning Point."

130 *many referred to the war as a "Double V"*: Delmont, "African-Americans Fighting Fascism and Racism."

130 *The exodus of men*: PBS, "Civil Rights"; Hartmann, *Home Front and Beyond*.

130 *Jobs opened for Mexican Americans*: Norton et al., *People and a Nation*, 746.

130 *When a couple suffers a betrayal*: Schwartz, "Esther Perel."

132 *was ground too shallow*: NASA, "Hubble Space Telescope."

133 *When they emerged the next morning*: "Theft That Made the 'Mona Lisa' a Masterpiece."

133 *At the time, the* Mona Lisa: Zug, "Stolen."

133 *But once it drew the attention*: "'LA GIOCONDA' IS STOLEN IN PARIS."

134 *When the Louvre reopened*: Kuper, "Who Stole the Mona Lisa?"

134 *Franz Kafka paid the empty wall*: Gekoski, "Fact-Check Fears."

134 *The following year, perhaps inspired*: Bernofsky, "On Translating Kafka's 'The Metamorphosis.'"

134 *The Louvre estimates*: Riding, "New Room with View of 'Mona Lisa.'"

134 *It now resides behind*: McKinney, "Mona Lisa Is Protected by a Fence."

136 *Among the few survivors*: Firestone, "Busting the Myths About Suicide."

136 *If we were to lose our planet*: Herbst, "Kevin Hines."

140 *"The biggest problem climate change poses"*: Scranton, "Learning How to Die."

143 *And I'm quite sure someone*: Parker, "Climate Migrants."

143 *"Listen to me," implores the soul*: "Dialogue of a Man with His Soul."

IV. DISPUTE WITH THE SOUL

150 *Climate change is a problem on the scale*: Griffin, "Carbon Majors Database."

150 *Just one hundred companies*: CDP, "100 Companies."

155 *He infiltrated a Nazi death camp*: Power, *"Problem from Hell."*

161 *Only 14 percent of Americans*: Leiserowitz et al., "Climate Change in the American Mind."

161 *significantly lower percentage*: Masci, "6 Facts About the Evolution Debate."

161 *the earth orbits the sun*: Neuman, "1 in 4 Americans."

161 *Sixty-nine percent of American voters*: Marlon et al., "Participation in the Paris Agreement."

164 *The richest 10 percent*: Oxfam, "Extreme Carbon Inequality."

164 *An estimated six million Bangladeshis*: Displacement Solutions, *Climate Displacement in Bangladesh.*

164 *Anticipated sea-level rises*: United Nations, "Statement by His Excellency Dr. Fakhruddin Ahmed."

165 *In 2018, it ranked Finland, Norway, and Denmark*: Helliwell et al., *World Happiness Report 2018.*

165 *The combined population of Finland*: Mapes, "Population of Nordic Countries."

165 *The average Bangladeshi is responsible*: "Carbon Emissions per Person, by Country."

165 *Bangladesh also happens to be*: Food and Agriculture Organization of the United Nations, "Annual Meat Consumption."

165 *In 2018, the average Finn*: Natural Resources Institute Finland, "What Was Eaten in Finland in 2016."

165 *Worldwide, more than 800 million*: World Food Program, "World Hunger Again on the Rise."

165 *More than 150 million*: Hunger Project, "Know Your World."

166 *One and a half million children*: United States Holocaust Memorial Museum, "Children During the Holocaust."

166 *Land that could feed*: Koneswaran and Nierenberg, "Global Farm Animal Production and Global Warming."

166 *The UN's former special rapporteur*: Ziegler, "Burning Food Crops."

166 *Twenty-three and a half million Americans live*: U.S. Department of Agriculture Economic Research Service, "Access to Affordable and Nutritious Food."

167 *It is true that a healthy traditional diet*: Caba, "Eating Healthy."

167 *But a healthy vegetarian diet*: Flynn and Schiff, "Economical Healthy Diets."

167 *For perspective, the median income*: FRED Economic Data, "Real Median Personal Income."

167 *There are fewer American farmers*: American Farm Bureau Federation, "Fast Facts About Agriculture"; "Farm Population Lowest Since 1850's"; United States Bureau of Labor Statistics, "Employment Projections Program."

169 *Livestock's contribution to environmental problems*: Steinfeld et al., *Livestock's Long Shadow.*

170 *In China, coal generates*: United States Energy Information Administration, "Chinese Coal-Fired Electricity Generation."

170 *How do we consider the fact*: Fischer and Keating, "How Eco-friendly Are Electric Cars?"

170 *And what about the other forms*: Wade, "Tesla's Electric Cars."

171 *The average U.S. and U.K. citizen*: Springmann et al., "Food System"; Carrington, "Reduction in Meat-Eating."

174 *Because of the fact that*: "Amazon Rainforest Deforestation 'Worst in 10 Years.'"

174 *Because American* CO_2e *emissions*: Plumer, "U.S. Carbon Emissions."

174 *Because of the 2017 discovery*: Wolf et al., "Global Livestock."

174 *and the 2018 discovery*: Pierre-Louis, "Ocean Warming."

174 *Because in the next thirty years*: Carrington, "Reduction in Meat-Eating."

V. MORE LIFE

187 *When Kevin Hines jumped*: Glionna, "Golden Gate's Suicide Lure."

187 *Armed conflicts will erupt*: Eckstein et al., "Global Climate Risk Index 2019."

190 *As one of the documents*: Shapiro and Epsztein, "Warsaw Ghetto Oyneg Shabes."

190 *Carved into the rock*: Croptrust, "Svalbard Global Seed Vault."

191 *Because the vault is kept*: Carrington, "Arctic Stronghold of World's Seeds."

191 *Another effort, called*: Frozen Ark Project, https://www.frozenark.org. Accessed February 1, 2019.

192 *It is written that "Noah"*: Genesis 9:13, https://biblehub.com/genesis/9-13.htm.

194 *Globally, more people die of suicide*: Dokoupil, "Why Suicide Has Become an Epidemic"; Lisa Schein, "More People Die from Suicide."

195 *A second note*: Parisienne et al., "Gay Rights Lawyer."

196 *Three months later*: Scranton, "Raising My Child."

196 *As David Wallace-Wells observes*: Wallace-Wells, "Uninhabitable Earth."

197 *He is referring to a paper*: Wynes and Nicholas, "Climate Mitigation Gap."

198 *In their book* Connected: Christakis and Fowler, *Connected.*

198 *"In a surprising regularity that"*: Christakis and Fowler, *Connected.*

199 *March on Washington for Jobs and Freedom*: Martin Luther King, Jr., Research and Education Institute, "March on Washington."

203 *"I can't protect my daughter"*: Scranton, "Raising My Child."

204 *A few months ago, a man*: Wilson, "His Body Was Behind the Wheel."

205 *"The birds chirped"*: Pilon, "I Found a Dead Body."

207 *When Stephen Hawking presided*: Low, "Cambridge Declaration on Consciousness."

209 *More than one hundred billion*: Kaneda and Haub, "How Many People Have Ever Lived on Earth?"

215 *While Armstrong prepared*: Safire, *Before the Fall*, 146.

215 *"In Event of Moon Disaster"*: A scanned image of the memo can be found at https://www.archives.gov/files/presidential-libraries/events/centennials/nixon/images/exhibit/rn100-6-1-2.pdf.

219 *The art curator João Ribas*: Paglen, *Last Pictures Project.*

221 *In Flannery O'Connor's story*: O'Connor, *Complete Stories*, 133.

APPENDIX: 14.5 PERCENT / 51 PERCENT

227 *"If this argument is right"*: Goodland and Anhang, "Livestock and Climate Change," 11.

227 *They recommend a 25 percent reduction*: Goodland and Anhang, "Response to 'Livestock and Greenhouse Gas Emissions.'"

228 *Those eager to challenge*: Lehmkuhl, "Livestock and Climate."

228 *After he retired*: Dr. Robert Goodland's personal website, accessed February 1, 2019, https://goodlandrobert.com; Goodland, "Robert Goodland Obituary."

228 *In a 2012 piece he wrote*: Goodland, "Meat Industry Pressure."

228 *"The FAO counts emissions"*: Goodland and Anhang, "Livestock and Climate Change," 13.

228 *In their 2010 follow-up*: Goodland and Anhang, "'Livestock and Climate Change': Critical Comments and Responses," 8.

229 *In addition, the FAO fails*: Goodland and Anhang, "Livestock and Climate Change," 14.

230 *The FAO also "uses citations"*: Ibid.

230 *"Emissions from livestock respiration"*: Steinfeld and Wassenaar, "Carbon and Nitrogen Cycles."

230 *"Today," Goodland and Anhang state*: Goodland and Anhang, "Livestock and Climate Change," 12.

230 *Goodland and Anhang add that*: Goodland and Anhang, "'Livestock and Climate Change': Critical Comments and Responses," 7.

230 *"Growth in markets"*: Goodland and Anhang, "Livestock and Climate Change," 13.

231 *"I emailed Mario"*: Lehmkuhl, "Livestock and Climate."

231 *Then, in 2012*: Goodland, "Meat Industry Pressure."

231 *Their stated objective*: Ibid.

232 *Goodland claims that this new partnership*: Steinfeld and Gerber, "Livestock Production."

232 *whereas the World Bank urges*: Goodland, "Meat Industry Pressure."

232 *Sure enough, in 2013*: Gerber et al., *Tackling Climate Change.*

232 *A 2014 UN General Assembly report*: Human Rights Council, "Report of the Special Rapporteur."

232 *The UNESCO authors write that*: Kanaly et al., "Meat Production," 10.

Bibliography

Allison, Tom. "Financial Health of Young America: Measuring Generational Declines Between Baby Boomers and Millennials." *Young Invincibles*, January 2017. https://younginvincibles.org/wp-content/uploads/2017/04/FHYA-Final2017-1-1.pdf.

"Amazon Rainforest Deforestation 'Worst in 10 Years,' Says Brazil." BBC News, November 24, 2018. https://www.bbc.com/news/world-latin-america-46327634.

American Chemical Society. "What Are the Greenhouse Gas Changes Since the Industrial Revolution?" Accessed January 31, 2019. https://www.acs.org/content/acs/en/climatescience/greenhousegases/industrialrevolution.html.

American Farm Bureau Federation. "Fast Facts About Agriculture." Accessed January 31, 2019. https://www.fb.org/newsroom/fast-facts.

American Merchant Marine at War. "Liberty Ship SS Robert E. Peary." Accessed January 30, 2019. www.usmm.org/peary.html.

Aton, Adam. "Earth Almost Certain to Warm by 2 Degrees Celsius." *Scientific American*, August 1, 2017. https://www.scientificamerican.com/article/earth-almost-certain-to-warm-by-2-degrees-celsius/.

Babaee, Samaneh, Ajay S. Nagpure, and Joseph F. DeCarolis. "How Much Do Electric Drive Vehicles Matter to Future U.S. Emissions?" *Environmental Science and Technology* 48 (2014): 1382–90. https://doi.org/10.1021/es4045677.

Ballew, Matthew, Jennifer Marlon, Xinran Wang, Anthony Leiserowitz, and Edward Maibach. "Importance of Global Warming as a Voting Issue in the U.S. Depends on Where People Live and What People Have Experienced." Yale Program on Climate Change Communication, November 2, 2018. climatecommunication.yale.edu/publications/climate-voters/.

Bar-On, Yinon M., Rob Phillips, and Ron Milo. "The Biomass Distribution on

Earth." *Proceedings of the National Academy of Sciences* 115, no. 25 (2018): 6506–11. https://doi.org/10.1073/pnas.1711842115.

Berger, Michele W. "Penn Psychologists Study Intense Awe Astronauts Feel Viewing Earth from Space." *Penn Today*, April 18, 2016. https://penntoday.upenn.edu/news/penn-psychologists-study-intense-awe-astronauts-feel-viewing-earth-space.

Bernofsky, Susan. "On Translating Kafka's 'The Metamorphosis.'" *New Yorker*, January 14, 2014. https://www.newyorker.com/books/page-turner/on-translating-kafkas-the-metamorphosis.

Bethge, Philip. "Urine Containers, 'Space Boots' and Artifacts Aren't Just Junk, Argue Archaeologists." *Spiegel Online*, March 18, 2010. www.spiegel.de/international/zeitgeist/saving-moon-trash-urine-containers-space-boots-and-artifacts-aren-t-just-junk-argue-archaeologists-a-684221.html.

Brahic, Catherine. "How Long Does It Take a Rainforest to Regenerate?" *New Scientist*, June 11, 2018. https://www.newscientist.com/article/dn14112-how-long-does-it-take-a-rainforest-to-regenerate/.

Brandt, Richard. "Google Divulges Numbers at I/O: 20 Billion Texts, 93 Million Selfies and More." *Silicon Valley Business Journal*, June 25, 2014. https://www.bizjournals.com/sanjose/news/2014/06/25/google-divulges-numbers-at-i-o-20-billion-texts-93.html.

British Nutrition Foundation. "How the War Changed Nutrition: From There to Now." Accessed January 12, 2019. https://www.nutrition.org.uk/nutritioninthenews/wartimefood/warnutrition.html.

Brody, Richard. "The Unicorn and 'The Karski Report.'" *New Yorker*. Accessed January 12, 2019. https://www.newyorker.com/culture/richard-brody/the-unicorn-and-the-karski-report.

Brookhaven National Laboratory. "The First Video Game?" Accessed January 30, 2019. https://www.bnl.gov/about/history/firstvideo.php.

Buehler, Jake. "This Tiny Fish Can Recognize Itself in a Mirror. Is It Self-Aware?" *National Geographic*, September 11, 2018. https://www.nationalgeographic.com/animals/2018/09/fish-cleaner-wrasse-self-aware-mirror-test-intelligence-news/.

Burke, Marshall, Solomon M. Hsiang, and Edward Miguel. "Climate and Conflict." *Annual Review of Economics* 7, no. 1 (2015): 577–617. https://web.stanford.edu/~mburke/papers/Burke%20Hsiang%20Miguel%202015.pdf.

Burkeman, Oliver. "We're All Climate Change Deniers at Heart." *Guardian*, June 8, 2015. https://www.theguardian.com/commentisfree/2015/jun/08/climate-change-deniers-g7-goal-fossil-fuels.

Caba, Justin. "Eating Healthy Could Get Costly: Healthy Diets Cost About $1.50 More Than Unhealthy Diets." *Medical Daily*, December 5, 2013. https://www.medicaldaily.com/eating-healthy-could-get-costly-healthy-diets-cost-about-150-more-unhealthy-diets-264432.

Calfas, Jennifer. "Americans Have So Much Debt They're Taking It to the Grave." *Money*, March 22, 2017. money.com/money/4709270/americans-die-in-debt/.

"Carbon Emissions per Person, by Country." *Guardian*, September 2, 2009. https://www.theguardian.com/environment/datablog/2009/sep/02/carbon-emissions-per-person-capita.

Carey, Benedict. "Alex, a Parrot Who Had a Way with Words, Dies." *New York Times*, September 10, 2007. https://www.nytimes.com/2007/09/10/science/10cnd-parrot.html.

Carrington, Damian. "Arctic Stronghold of World's Seeds Flooded After Permafrost Melts." *Guardian*, May 19, 2017. https://www.theguardian.com/environment/2017/may/19/arctic-stronghold-of-worlds-seeds-flooded-after-permafrost-melts.

———. "Huge Reduction in Meat-Eating 'Essential' to Avoid Climate Breakdown." *Guardian*, October 10, 2018. https://www.theguardian.com/environment/2018/oct/10/huge-reduction-in-meat-eating-essential-to-avoid-climate-breakdown.

———. "Humans Just 0.01% of All Life but Have Destroyed 83% of Wild Mammals—Study." *Guardian*, May 21, 2018. https://www.theguardian.com/environment/2018/may/21/human-race-just-001-of-all-life-but-has-destroyed-over-80-of-wild-mammals-study.

CDP. "New Report Shows Just 100 Companies Are Source of Over 70% of Emissions." July 10, 2017. https://www.cdp.net/en/articles/media/new-report-shows-just-100-companies-are-source-of-over-70-of-emissions.

Center for Sustainable Systems. "Carbon Footprint Factsheet." University of Michigan, 2018. Accessed March 10, 2019. css.umich.edu/factsheets/carbon-footprint-factsheet.

Centers for Disease Control and Prevention. "Tobacco-Related Mortality." January 17, 2018. https://www.cdc.gov/tobacco/data_statistics/fact_sheets/health_effects/tobacco_related_mortality/index.htm.

Chase, Robin. "Car-Sharing Offers Convenience, Saves Money and Helps the Environment." United States Department of State, Bureau of International Information Programs. Accessed February 5, 2019. https://photos.state.gov/libraries/cambodia/30486/Publications/everyone_in_america_own_a_car.pdf.

Christakis, Nicholas A., and James H. Fowler. *Connected: The Surprising Power of Our Social Networks and How They Shape Our Lives*. New York: Little, Brown, 2009.

Clark, Duncan. "How Long Do Greenhouse Gases Stay in the Air?" *Guardian*, January 16, 2012. Published in conjunction with Carbon Brief. https://www.theguardian.com/environment/2012/jan/16/greenhouse-gases-remain-air.

Climate and Land Use Alliance. "Five Reasons the Earth's Climate Depends on Forests." Accessed May 10, 2018. http://www.climateandlandusealliance.org/scientists-statement/.

Collingham, Lizzie. *The Taste of War: World War II and the Battle for Food*. New York: Penguin, 2013.

Compassion in World Farming. *Strategic Plan 2013–2017*. Accessed January 30, 2019. https://www.ciwf.org.uk/media/3640540/ciwf_strategic_plan_20132017.pdf.

Conca, James. "Children Change the Climate in the U.S. Supreme Court—1st Climate Lawsuit Goes Forward." *Forbes*, August 3, 2018. https://www.forbes.com/sites/jamesconca/2018/08/03/children-change-the-climate-in-the-us-supreme-court-1st-climate-lawsuit-goes-forward/#1b34f8e53547.

Coren, Michael J. "Humans Have Depleted the Earth's Natural Resources with Five Months Still to Go in 2018." *Quartz*, August 1, 2018. https://qz.com/1345205/humans-have-depleted-the-earths-natural-resources-with-five-months-still-to-go-in-2018/.

Croptrust. "Svalbard Global Seed Vault." Accessed January 25, 2019. https://www.croptrust.org/our-work/svalbard-global-seed-vault/.

Cushman, John H., Jr. "Exxon Made Deep Cuts in Climate Research Budget in the 1980s." *Inside Climate News*, November 25, 2015. https://insideclimatenews.org/news/25112015/exxon-deep-cuts-climate-change-research-budget-1980s-global-warming.

Daggett, Stephen. *Costs of Major U.S. Wars.* U.S. Library of Congress, Congressional Research Service, RS22926, 2010.

Dahl, Melissa. "Why Can't You Smell Your Own Home?" *The Cut*, August 26, 2014. https://www.thecut.com/2014/08/why-cant-you-smell-your-own-home.html.

Davidai, Shai, Thomas Gilovich, and Lee D. Ross. "The Meaning of Default Options for Potential Organ Donors." *Proceedings of the National Academy of Sciences*, 2012: 15201–205. https://stanford.app.box.com/s/yohfziywajw3nmwxo7d3ammndihibe7g.

"Deforestation and Its Extreme Effect on Global Warming." *Scientific American*. Accessed January 31, 2019. https://www.scientificamerican.com/article/deforestation-and-global-warming/.

Delmont, Matthew. "African-Americans Fighting Fascism and Racism, from WWII to Charlottesville." *The Conversation*, August 21, 2017. https://theconversation.com/african-americans-fighting-fascism-and-racism-from-wwii-to-charlottesville-82551.

Delta Dental. *2014 Oral Health and Well-Being Survey*. 2014. https://www.deltadentalnj.com/employers/downloads/DDPAOralHealthandWellBeingSurvey.pdf.

Dennis, Brady. "Who Still Smokes in the United States—in Seven Simple Charts." *Washington Post*, November 12, 2015. https://www.washingtonpost.com/news/to-your-health/wp/2015/11/12/smoking-among-u-s-adults-has-fallen-to-historic-lows-these-7-charts-show-who-still-lights-up-the-most/.

"Dialogue of a Man with His Soul." Ethics of Suicide Digital Archive, University of Utah. Accessed February 5, 2019. https://ethicsofsuicide.lib.utah.edu/selections/egyptian-didactic-tale/.

Di Leberto, Tom. "Summer Heat Wave Arrives in Europe." Climate.gov, July 14, 2015. https://www.climate.gov/news-features/event-tracker/summer-heat-wave-arrives-europe.

Displacement Solutions. *Climate Displacement in Bangladesh: The Need for Ur-*

gent Housing, Land and Property (HLT) Rights Solutions. May 2012. https://unfccc.int/files/adaptation/groups_committees/loss_and_damage_executive_committee/application/pdf/ds_bangladesh_report.pdf.

Dohmen, Thomas J. "In Support of the Supporters? Do Social Forces Shape Decisions of the Impartial?" Institute for the Study of Labor (IZA) Bonn, April 2003. http://ftp.iza.org/dp755.pdf.

Dokoupil, Tony. "Why Suicide Has Become an Epidemic—and What We Can Do to Help." *Newsweek*, May 23, 2013. https://www.newsweek.com/2013/05/22/why-suicide-has-become-epidemic-and-what-we-can-do-help-237434.html.

Dunn, Marcia. "SpaceX Chief Envisions 1,000 Passenger Ships Flying to Mars." AP, September 27, 2016. https://apnews.com/a8c262f520c14ee583fdbb07d1f82a25.

Durisin, Megan, and Shruti Date Singh. "Americans' Meat Consumption Set to Hit a Record in 2018." *Seattle Times*, January 2, 2018. https://www.seattletimes.com/business/americans-meat-consumption-set-to-hit-a-record-in-2018/.

Eberle, Jaelyn J., Henry C. Fricke, John D. Humphrey, Logan Hackett, Michael G. Newbrey, and J. Howard Hutchison. "Seasonal Variability in Arctic Temperatures During Early Eocene Time." *Earth and Planetary Science Letters* 296, no. 3–4 (August 2010): 481–86. https://doi.org/10.1016/j.epsl.2010.06.005.

Eckstein, David, Marie-Lena Hutfils, and Maik Wings. "Global Climate Risk Index 2019: Who Suffers Most from Extreme Weather Events? Weather-Related Loss Events in 2017 and 1998 to 2017." *Germanwatch*, December 2018. https://www.germanwatch.org/sites/germanwatch.org/files/Global%20Climate%20Risk%20Index%202019_2.pdf.

Erb, Karl-Heinz, Thomas Kastner, Christoph Plutzar, Anna Liza S. Bais, Nuno Carvalhais, Tamara Fetzel, Simone Gingrich, Helmut Haberl, Christian Lauk, Maria Niedertscheider, Julia Pongratz, Martin Thurner, and Sebastiaan Luyssaert. "Unexpectedly Large Impact of Forest Management and Grazing on Global Vegetation Biomass." *Nature* 553, no. 7686 (January 2018). https://www.nature.com/articles/nature25138.

Erman, Adolf. *The Ancient Egyptians: A Sourcebook of Their Writings.* Translated by Ayward M. Blackman. New York: Harper and Row, 1966.

"Farm Population Lowest Since 1850's." *New York Times*, July 20, 1988. https://www.nytimes.com/1988/07/20/us/farm-population-lowest-since-1850-s.html.

Ferreira, Becky. "Seeing Earth from Space Is the Key to Saving Our Species from Itself." *Motherboard*, October 12, 2016. https://motherboard.vice.com/en_us/article/bmvpxq/to-save-humanity-look-at-earth-from-space-overview-effect.

Firestone, Lisa. "Busting the Myths About Suicide." *PsychAlive.* Accessed January 24, 2019. https://www.psychalive.org/busting-the-myths-about-suicide/.

Fischer, Hilke, and Dave Keating. "How Eco-friendly Are Electric Cars?" Deutsche Welle, April 8, 2017. https://www.dw.com/en/how-eco-friendly-are-electric-cars/a-19441437.

Fishcount. "Number of Farmed Fish Slaughtered Each Year." Accessed January 30,

2019. fishcount.org.uk/fish-count-estimates-2/numbers-of-farmed-fish-slaughtered-each-year.

Fitch, Riley. "Julia Child: The OSS Years." *Wall Street Journal*, August 19, 2008. https://www.wsj.com/articles/SB121910345904851347.

Flynn, Mary M., and Andrew R. Schiff. "Economical Healthy Diets (2012): Including Lean Animal Protein Costs More Than Using Extra Virgin Olive Oil." *Journal of Hunger and Environmental Nutrition* 10, no. 4 (2015): 467–82. https://doi.org/10.1080/19320248.2015.1045675.

Food and Agriculture Organization of the United Nations. "Current Worldwide Annual Meat Consumption per Capita, Livestock and Fish Primary Equivalent." Accessed January 26, 2019. http://faostat.fao.org/site/610/DesktopDefault.aspx?PageID=610#ancor.

———. "Deforestation Causes Global Warming." FAO Newsroom, September 4, 2006. www.fao.org/newsroom/en/news/2006/1000385/index.html.

Food Will Win the War. Short film. Walt Disney Productions. YouTube, April 24, 2012. https://www.youtube.com/watch?v=HeTqKKCm3Tg.

FRED Economic Data. "Real Median Personal Income in the United States." Federal Reserve Bank of St. Louis. Accessed February 1, 2019. https://fred.stlouisfed.org/series/MEPAINUSA672N.

Frischmann, Chad. "100 Solutions to Reverse Global Warming." TED Talk video. YouTube, December 19, 2018. https://www.youtube.com/watch?v=D4vjGSiRGKY&feature=youtu.be.

Gallagher, Hugh Gregory. *FDR's Splendid Deception: The Moving Story of Roosevelt's Massive Disability—and the Intense Efforts to Conceal It from the Public.* St. Petersburg, FL: Vandamere Press, 1999.

Gannon, Michael. *Operation Drumbeat: The Dramatic True Story of Germany's First U-Boat Attacks Along the American Coast in World War II.* New York: Harper and Row, 1996.

Ganzel, Bill. "Shrinking Farm Numbers." Wessels Living History Farm, 2007. https://livinghistoryfarm.org/farminginthe50s/life_11.html.

Garan, Ron. *The Orbital Perspective: Lessons in Seeing the Big Picture from a Journey of 71 Million Miles.* Oakland, CA: Berrett-Koehler, 2015.

Gates, Bill. "Climate Change and the 75% Problem." *Gatesnotes* (blog), October 17, 2018. https://www.gatesnotes.com/Energy/My-plan-for-fighting-climate-change.

Gekoski, Rick. "Fact-Check Fears." *Guardian*, September 9, 2011. https://www.theguardian.com/books/2011/sep/09/fact-check-rick-gekoski.

George C. Marshall Foundation. "National Nutrition Month and Rationing." March 4, 2016. https://www.marshallfoundation.org/blog/national-nutrition-month-rationing/.

Gerbens-Leenes, P. W., M. M. Mekonnen, and A. Y. Hoekstra. "The Water Footprint of Poultry, Pork and Beef: A Comparative Study in Different Countries and Production Systems." *Water Resources and Industry* 1–2 (2013): 25–36. https://doi.org/10.1016/j.wri.2013.03.001.

Gerber, P. J., H. Steinfeld, B. Henderson, A. Mottet, C. Opio, J. Dijkman, A. Falcucci, and G. Tempio. *Tackling Climate Change Through Livestock: A Global Assessment of Emissions and Mitigation Opportunities.* Food and Agriculture Organization of the United Nations, Rome, 2013. http://www.fao.org/3/a-i3437e.pdf.

Ghosh, Amitav. *The Great Derangement: Climate Change and the Unthinkable.* Chicago: University of Chicago Press, 2016.

Girod, B., D. P. van Vuuren, and E. G. Hertwich. "Climate Policy Through Changing Consumption Choices: Options and Obstacles for Reducing Greenhouse Gas Emissions." *Global Environmental Change* 25 (2014): 5–15.

Glionna, John M. "Survivor Battles Golden Gate's Suicide Lure." *Seattle Times*, June 4, 2005. https://www.seattletimes.com/nation-world/survivor-battles-golden-gates-suicide-lure/.

"Global Carbon Dioxide Emissions Rose Almost 3% in 2018." CBC. December 5, 2018. https://www.cbc.ca/news/technology/carbon-pollution-increase-1.4934096.

Goldhill, Olivia. "Astronauts Report an 'Overview Effect' from the Awe of Space Travel—and You Can Replicate It Here on Earth." *Quartz*, September 6, 2015. https://qz.com/496201/astronauts-report-an-overview-effect-from-the-awe-of-space-travel-and-you-can-replicate-it-here-on-earth/.

Goodland, Robert. "FAO Yields to Meat Industry Pressure on Climate Change." *New York Times*, July 11, 2012. https://bittman.blogs.nytimes.com/2012/07/11/fao-yields-to-meat-industry-pressure-on-climate-change/.

Goodland, Robert, and Jeff Anhang. "Livestock and Climate Change." *World Watch Magazine*, November–December 2009, 10–19. http://www.worldwatch.org/files/pdf/Livestock%20and%20Climate%20Change.pdf.

———. "'Livestock and Climate Change': Critical Comments and Responses." *World Watch Magazine*, March–April 2010, 7–9. www.chompingclimatechange.org/wp-content/uploads/2015/01/Livestock-and-Climate-Change-critical-comments-and-responses.pdf.

———. "Response to 'Livestock and Greenhouse Gas Emissions: The Importance of Getting the Numbers Right,' by Herrero et al. [Anim. Feed Sci. Technol. 166–167: 779–782]." *Animal Feed Science and Technology* 172, 252–56. https://www.sciencedirect.com/science/article/pii/S0377840111005177.

Goodland, Tom. "Robert Goodland Obituary." *Guardian*, February 4, 2014. https://www.theguardian.com/environment/2014/feb/05/robert-goodland.

Gorman, James. "It Could Be the Age of the Chicken, Geologically." *New York Times*, December 11, 2018. https://www.nytimes.com/2018/12/11/science/chicken-anthropocene-archaeology.html.

Greshko, Michael, Laura Parker, Brian Clark Howard, Daniel Stone, Alejandra Borunda, and Sarah Gibbens. "How Trump Is Changing the Environment."

National Geographic, January 17, 2019. https://news.nationalgeographic.com/2017/03/how-trump-is-changing-science-environment/.

Griffin, Paul. "The Carbon Majors Database: CDP Carbon Majors Report 2017." CDP, July 2017. http://climateaccountability.org/pdf/CarbonMajorsRpt2017%20Jul17.pdf.

Guggenheim, Davis, dir. *An Inconvenient Truth*. 2006: Paramount.

Harris Poll. "Carrying On Tradition Around the Thanksgiving Table." Accessed January 30, 2019. https://theharrispoll.com/thanksgiving-is-just-around-the-corner-and-americans-across-the-country-are-planning-what-to-serve-who-theyll-dine-with-and-where-theyll-eat-a-vast-majority-of-adults-indicate-the/.

Hartmann, Susan M. *The Home Front and Beyond: American Women in the 1940s*. Boston: Twayne, 1982.

Harvey, Fiona. "World Has Three Years Left to Stop Dangerous Climate Change, Warn Experts." *Guardian*, June 28, 2017. https://www.theguardian.com/environment/2017/jun/28/world-has-three-years-left-to-stop-dangerous-climate-change-warn-experts.

Helliwell, John F., Richard Layard, and Jeffrey D. Sachs. *World Happiness Report 2018*. Accessed February 1, 2019. https://s3.amazonaws.com/happiness-report/2018/WHR_web.pdf.

Herbst, Diane, "Kevin Hines Survived a Jump Off the Golden Gate Bridge. Now, He's Helping Others Avoid Suicide." PSYCOM, June 8, 2018, https://www.psycom.net/kevin-hines-survived-golden-gate-bridge-suicide/.

Hershfield, Hal. "You Make Better Decisions If You 'See' Your Senior Self." *Harvard Business Review*, June 2013. https://hbr.org/2013/06/you-make-better-decisions-if-you-see-your-senior-self.

Hoekstra, Arjen Y., and Mesfin M. Mekonnen. "The Water Footprint of Humanity." *Proceedings of the National Academy of Sciences* 109, no. 9 (February 2012), 3232–37. https://doi.org/10.1073/pnas.1109936109.

Holford, Thomas R., Rafael Meza, Kenneth E. Warner, Clare Meernik, Jihyoun Jeon, Suresh H. Moolgavkar, and David T. Levy. "Tobacco Control and the Reduction in Smoking-Related Premature Deaths in the United States, 1964–2012." *JAMA* 311, no. 2 (2014): 164–71. https://doi.org/10.1001/jama.2013.285112.

Huicochea, Alexis. "Man Lifts Car Off Pinned Cyclist." *Arizona Daily Star*, July 28, 2006. https://tucson.com/news/local/crime/man-lifts-car-off-pinned-cyclist/article_e7f04bbd-309b-5c7e-808d-1907d91517ac.html.

Human Rights Council. "Report of the Special Rapporteur on the Right to Food, Olivier De Schutter." United Nations General Assembly, January 24, 2014. www.srfood.org/images/stories/pdf/officialreports/20140310_finalreport_en.pdf.

Hunger Project. "Know Your World: Facts About Hunger and Poverty." November 2017. https://www.thp.org/knowledge-center/know-your-world-facts-about-hunger-poverty/.

Institute for Operations Research and the Management Sciences. "Carefully Chosen Wording Can Increase Donations by Over 300 Percent." *ScienceDaily*, November 8, 2016. www.sciencedaily.com/releases/2016/11/161108120317.htm.

Institute of Noetic Sciences. "Our Story." IONS EarthRise. Accessed January 31, 2019. https://noetic.org/earthrise/about/overview.

Intergovernmental Panel on Climate Change. *Climate Change 2013: The Physical Science Basis—Contribution of Working Group I to the Fifth Assessment Report of the Intergovernmental Panel on Climate Change*, chap. 8 (Cambridge, UK, and New York: Cambridge University Press, 2013), 711–14, table 8.7. https://doi.org/10.1017/CBO9781107415324.

———. *Global Warming of 1.5°C: An IPCC Special Report on the Impacts of Global Warming of 1.5°C Above Pre-industrial Levels and Related Global Greenhouse Gas Emission Pathways, in the Context of Strengthening the Global Response to the Threat of Climate Change, Sustainable Development, and Efforts to Eradicate Poverty.* Edited by V. Masson-Delmotte, P. Zhai, H. O. Pörtner, D. Roberts, J. Skea, P. R. Shukla, A. Pirani, W. Moufouma-Okia, C. Péan, R. Pidcock, S. Connors, J.B.R. Matthews, Y. Chen, X. Zhou, M. I. Gomis, E. Lonnoy, T. Maycock, M. Tignor, and T. Waterfield. 2018.

Jacobson, Mark Z., and Mark A. Delucchi. "A Path to Sustainable Energy by 2030." *Scientific American*, November 2009, 58–65. https://web.stanford.edu/group/efmh/jacobson/Articles/I/sad1109Jaco5p.indd.pdf.

Jardine, Phil. "Patterns in Palaeontology: The Paleocene–Eocene Thermal Maximum." *Palaeontology Online* 1, no. 5. Accessed January 31, 2019. https://www.palaeontologyonline.com/articles/2011/the-paleocene-eocene-thermal-maximum/.

Joint Study for the Atmosphere and the Ocean. "PDO Index." University of Washington. http://research.jisao.washington.edu/pdo/PDO.latest.

Jouzel, Jean, Valérie Masson-Delmotte, Olivier Cattani, Gabrielle B. Dreyfus, Sonia Falourd, Georg Hoffmann, Benedicte Minster, Julius Nouet, J. M. Barnola, Jérôme Chappellaz, Hubertus Fischer, Jean Charles Gallet, S.E.J. Johnsen, Markus Leuenberger, Laetitia Loulergue, Dieter Lüthi, Hans Oerter, Frédéric Parrenin, Grant M. Raisbeck, Dominique Raynaud, Adrian Schilt, Jakob Schwander, Enricomaria Selmo, Roland A. Souchez, Renato Spahni, Bernhard Stauffer, Jorgen Peder Steffensen, Barbara Stenni, Thomas F. Stocker, J. L. Tison, Maria Werner, and Eric W. Wolff. "Orbital and Millennial Antarctic Climate Variability over the Past 800,000 Years." *Science* 317, no. 5839 (2007): 793–96. https://doi.org/10.1126/science.1141038.

Kanaly, Robert A., Lea Ivy O. Manzanero, Gerard Foley, Sivanandam Panneerselvam,

Darryl Macer. "Energy Flow, Environment and Ethical Implications for Meat Production." Ethics and Climate Change in Asia and the Pacific (ECCAP) Project, 2010. www.eubios.info/yahoo_site_admin/assets/docs/ECCAPWG13.83161418.pdf.

Kaneda, Toshiko, and Carl Haub. "How Many People Have Ever Lived on Earth?" Population Reference Bureau, March 9, 2018. https://www.prb.org/howmanypeoplehaveeverlivedonearth/.

Kastberger, Gerald, Evelyn Schmelzer, and Ilse Kranner. "Social Waves in Giant Honeybees Repel Hornets." *PLOS ONE* 3, no. 9 (2008): e3141. https://doi.org/10.1371/journal.pone.0003141.

Kearl, Michael C. *Endings: A Sociology of Death and Dying.* New York: Oxford University Press, 1989.

Khan, Sammyh S., Nick Hopkins, Stephen Reicher, Shruti Tewari, Narayanan Srinivasan, and Clifford Stevenson. "How Collective Participation Impacts Social Identity: A Longitudinal Study from India." *Political Psychology* 37 (2016): 309–25. https://doi.org/10.1111/pops.12260.

Kim, B. F., R. E. Santo, A. P. Scatterday, J. P. Fry, C. M. Synk, S. R. Cebron, M. M. Mekonnen, A. Y. Hoekstra, S. de Pee, M. W. Bloem, R. A. Neff, and K. E. Nachman. "Country-Specific Dietary Shifts to Mitigate Climate and Water Crises." Publication pending.

Kim, Brent, Roni Neff, Raychel Santo, and Juliana Vigorito. "The Importance of Reducing Animal Product Consumption and Wasted Food in Mitigating Catastrophic Climate Change." Johns Hopkins Center for a Livable Future, 2015. https://www.jhsph.edu/research/centers-and-institutes/johns-hopkins-center-for-a-livable-future/_pdf/research/clf_reports/importance-of-reducing-animal-product-consumption-and-wasted-food-in-mitigating-catastrophic-climate-change.pdf.

Kluger, Jeffrey. "Earth from Above: The Blue Marble." *Time*, February 9, 2012. time.com/3785942/blue-marble/.

Knapton, Sarah. "Human Race Is Doomed If We Do Not Colonize the Moon and Mars, Says Stephen Hawking." *Telegraph*, June 20, 2017. https://www.telegraph.co.uk/science/2017/06/20/human-race-doomed-do-not-colonise-moon-mars-says-stephen-hawking/.

Koneswaran, G., and D. Nierenberg. "Global Farm Animal Production and Global Warming: Impacting and Mitigating Climate Change." *Environmental Health Perspectives* 116, no. 5 (2008): 578–82. https://www.ncbi.nlm.nih.gov/pmc/articles/PMC2367646/.

Kotecki, Peter. "Jeff Bezos Is the Richest Man in Modern History—Here's How He Spends on Philanthropy." *Business Insider*, September 13, 2018. https://www.businessinsider.com/jeff-bezos-richest-person-modern-history-spends-on-charity-2018-7.

Kuper, Simon. "Who Stole the Mona Lisa?" *Slate*, August 7, 2011. https://slate.com

/human-interest/2011/08/who-stole-the-mona-lisa-the-world-s-most-famous-art-heist-100-years-on.html.

"'LA GIOCONDA' IS STOLEN IN PARIS; Masterpiece of Lionardo da Vinci Vanishes from Louvre [. . .]." *New York Times*, August 23, 1911. https://www.nytimes.com/1911/08/23/archives/la-gioconda-is-stolen-in-paris-masterpiece-of-lionardo-da-vinci.html.

Lamble, Lucy. "With 250 Babies Born Each Minute, How Many People Can the Earth Sustain?" *Guardian*, April 23, 2018. https://www.theguardian.com/global-development/2018/apr/23/population-how-many-people-can-the-earth-sustain-lucy-lamble.

Lavelle, Marianne. "2016: Obama's Climate Legacy Marked by Triumphs and Lost Opportunities." *Inside Climate News*, December 26, 2016. https://insideclimatenews.org/news/23122016/obama-climate-change-legacy-trump-policies.

Lee, Bruce Y. "What Is 'Selfitis' and When Does Taking Selfies Become a Real Problem?" *Forbes*, December 26, 2017. https://www.forbes.com/sites/brucelee/2017/12/26/what-is-selfitis-and-when-does-taking-selfies-become-a-real-problem/#648994c330dc.

Lehmkuhl, Vance. "Livestock and Climate: Whose Numbers Are More Credible?" *Philadelphia Inquirer*, March 2, 2012. https://www.philly.com/philly/blogs/earth-to-philly/Livestock-and-climate-Whose-numbers-are-more-credible.html.

Leiserowitz, Anthony, Edward Maibach, Seth Rosenthal, John Kotcher, Matthew Ballew, Matthew Goldberg, and Abel Gustafson. "Climate Change in the American Mind: December 2018." Yale Program on Climate Change Communication, January 22, 2019. climatecommunication.yale.edu/publications/climate-change-in-the-american-mind-december-2018/2/.

Leskin, Paige. "The 13 Tech Billionaires Who Donate the Biggest Percentage of Their Wealth to Charity." *Business Insider*, January 31, 2019. https://www.businessinsider.com/tech-billionaires-who-donate-most-to-charity-2019-1.

Levine, Morgan E., Jorge A. Suarez, Sebastian Brandhorst, Priya Balasubramanian, Chia-Wei Cheng, Federica Madia, Luigi Fontana, Mario G. Mirisola, Jaime Guevara-Aguirre, Junxiang Wan, Giuseppe Passarino, Brian K. Kennedy, Min Wei, Pinchas Cohen, Eileen M. Crimmins, and Valter D. Longo. "Low Protein Intake Is Associated with a Major Reduction in IGF-1, Cancer, and Overall Mortality in the 65 and Younger but Not Older Population." *Cell Metabolism* 19 (2014): 407–17. https://doi.org/10.1016/j.cmet.2014.02.006.

Lewin, Zofia, and Wladyslaw Bartoszewski, eds. *Righteous Among Nations: How the Poles Helped the Jews 1939–1945*. London: Earlscourt 42 Publications, 1969. www.writing.upenn.edu/~afilreis/Holocaust/karski.html.

Lincoln, Abraham. "Proclamation of Thanksgiving." Accessed January 30, 2019. www.abrahamlincolnonline.org/lincoln/speeches/thanks.htm.

Low, Philip. "The Cambridge Declaration on Consciousness." Presented at the Francis Crick Memorial Conference on Consciousness in Human and Non-

Human Animals, Churchill College, University of Cambridge, July 7, 2012. fcmconference.org/img/CambridgeDeclarationOnConsciousness.pdf.

Ma, Qiancheng. "Greenhouse Gases: Refining the Role of Carbon Dioxide." National Aeronautics and Space Administration, Goddard Institute for Space Studies, March 1998. https://www.giss.nasa.gov/research/briefs/ma_01/.

Mann, Michael E., and Lee R. Kump. *Dire Predictions: The Visual Guide to the Findings of the IPCC.* 2nd ed. London: DK, 2015.

Mapes, Terri. "The Population of Nordic Countries." *TripSavvy*, August 17, 2018. https://www.tripsavvy.com/population-in-nordic-countries-1626872.

Margulis, Sergio. "Causes of Deforestation of the Brazilian Amazon." World Bank Working Paper 22, 2004. http://documents.worldbank.org/curated/en/758171468768828889/pdf/277150PAPER0wbwp0no1022.pdf.

Marlon, Jennifer, Eric Fine, and Anthony Leiserowitz. "Majorities of Americans in Every State Support Participation in the Paris Agreement." Yale Program on Climate Change Communication, May 8, 2017. climatecommunication.yale.edu/publications/paris_agreement_by_state/.

Marshall, George. *Don't Even Think About It: Why Our Brains Are Wired to Ignore Climate Change.* New York: Bloomsbury USA, 2014.

Martin Luther King, Jr., Research and Education Institute. "March on Washington for Jobs and Freedom." Stanford University. Accessed January 25, 2019. https://kinginstitute.stanford.edu/encyclopedia/march-washington-jobs-and-freedom.

Masci, David. "For Darwin Day, 6 Facts About the Evolution Debate." Pew Research Center, February 10, 2017. www.pewresearch.org/fact-tank/2017/02/10/darwin-day/.

Matthews, Dylan. "Donald Trump Has Tweeted Climate Change Skepticism 115 Times. Here's All of It." *Vox*, June 1, 2017. https://www.vox.com/policy-and-politics/2017/6/1/15726472/trump-tweets-global-warming-paris-climate-agreement.

McDonald, Charlotte. "How Many Earths Do We Need?" BBC News, June 16, 2015. https://www.bbc.com/news/magazine-33133712.

McKibben, Bill. "Global Warming's Terrifying New Math." *Rolling Stone*, July 19, 2012. https://www.rollingstone.com/politics/politics-news/global-warmings-terrifying-new-math-188550/.

———. "Up Against Big Oil in the Midterms." *New York Times*, November 7, 2018. https://www.nytimes.com/2018/11/07/opinion/climate-midterms-emissions-fossil-fuels.html.

McKie, Robin. "A Jab for Elvis Helped Beat Polio. Now Doctors Have Recruited Him Again." *Guardian*, April 23, 2016. https://www.theguardian.com/society/2016/apr/24/elvis-presley-polio-vaccine-world-immunisation-week.

McKinney, Kelsey. "The Mona Lisa Is Protected by a Fence that Beyoncé and Jay-Z Ignored." *Vox*, October 13, 2014. https://www.vox.com/xpress/2014/10

/13/6969099/the-mona-lisa-is-protected-by-a-fence-that-beyonce-and-jay-z-ignored.

Meixler, Eli. "Half of All Wildlife Could Disappear from the Amazon, Galapagos and Madagascar Due to Climate Change." *Time*, March 14, 2018. time.com/5198732/wwf-climate-change-report-wildlife/.

Mellström, Carl, and Magnus Johannesson. "Crowding Out in Blood Donation: Was Titmuss Right?" *Journal of the European Economic Association* 6 (2008): 845–63. https://doi.org/10.1162/JEEA.2008.6.4.845.

Milman, Oliver. "Climate Change Is Making Hurricanes Even More Destructive, Research Finds." *Guardian*, November 14, 2018. https://www.theguardian.com/environment/2018/nov/14/climate-change-hurricanes-study-global-warming.

Mooallem, Jon. "The Last Supper: Living by One-Handed Food Alone." *Harper's Magazine*, July 2005.

Moore, David W. "Nine of Ten Americans View Smoking as Harmful." Gallup News Service, October 7, 1999. https://news.gallup.com/poll/3553/nine-ten-americans-view-smoking-harmful.aspx.

Mortimer, Ian. "The Mirror Effect: How the Rise of Mirrors in the Fifteenth Century Shaped Our Idea of the Individual." *Lapham's Quarterly*, November 9, 2016. https://www.laphamsquarterly.org/roundtable/mirror-effect.

Morton, Ella. "Object of Intrigue: The Dummy Paratroopers of WWII." Atlas Obscura, November 10, 2015. https://www.atlasobscura.com/articles/object-of-intrigue-the-dummy-paratroopers-of-wwii.

Moss, Michael. "Nudged to the Produce Aisle by a Look in the Mirror." *New York Times*, August 27, 2013. https://www.nytimes.com/2013/08/28/dining/wooing-us-down-the-produce-aisle.html.

Muise, Amy, Elaine Giang, and Emily A. Impett. "Post Sex Affectionate Exchanges Promote Sexual and Relationship Satisfaction." *Archives of Sexual Behavior* 47, no. 3 (October 2014): 1391–1402. https://doi.org/10.1007/s10508-014-0305-3.

Nardo, Don. *The Blue Marble: How a Photograph Revealed Earth's Fragile Beauty.* Mankato, MN: Compass Point, 2014.

NASA. "The Hubble Space Telescope." Goddard Space Flight Center. Accessed February 1, 2019. https://asd.gsfc.nasa.gov/archive/hubble/missions/sm1.html.

———. "Mars Facts." Accessed February 1, 2019. https://mars.nasa.gov/allaboutmars/facts/#?c=inspace&s=distance.

———. "Scientific Consensus: Earth's Climate Is Warming." Accessed January 30, 2019. https://climate.nasa.gov/scientific-consensus/.

NASA Earth Observatory. "How Is Today's Warming Different from the Past?" June 3, 2010. https://earthobservatory.nasa.gov/features/GlobalWarming/page3.php.

———. "Is Current Warming Natural?" June 3, 2010. https://www.earthobservatory.nasa.gov/features/GlobalWarming/page4.php.

National Center for Injury Prevention and Control. "Suicide Rising Across the US." Centers for Disease Control and Prevention. Accessed January 25, 2019. https://www.cdc.gov/vitalsigns/suicide/index.html.

National Centers for Environmental Information. "Glacial-Interglacial Cycles." National Oceanic and Atmospheric Administration. Accessed January 31, 2019. https://www.ncdc.noaa.gov/abrupt-climate-change/Glacial-Interglacial%20Cycles.

———. "Global Climate Report—Annual 2017." National Oceanic and Atmospheric Administration. Accessed January 31, 2019. https://www.ncdc.noaa.gov/sotc/global/201713.

National Research Council. *Climate Intervention: Reflecting Sunlight to Cool Earth.* Washington, D.C.: National Academies Press, 2015. https://doi.org/10.17226/18988.

National Snow and Ice Data Center. "All About Sea Ice: Albedo." Accessed January 30, 2019. https://nsidc.org/cryosphere/seaice/processes/albedo.html.

National Weather Service Climate Prediction Center. "Cold and Warm Episodes by Season." http://origin.cpc.ncep.noaa.gov/products/analysis_monitoring/ensostuff/ONI_v5.php.

Natural Resources Institute Finland. "What Was Eaten in Finland in 2016." June 29, 2017. https://www.luke.fi/en/news/eaten-finland-2016/.

Neuman, Scott. "1 in 4 Americans Thinks the Sun Goes Around the Earth, Survey Says." NPR, February 14, 2014. https://www.npr.org/sections/thetwo-way/2014/02/14/277058739/1-in-4-americans-think-the-sun-goes-around-the-earth-survey-says.

New Mexico Museum of Space History. "International Space Hall of Fame: William A. Anders." Accessed January 24, 2019. http://www.nmspacemuseum.org/halloffame/detail.php?id=71.

New York University. "Scientists Find Evidence That Siberian Volcanic Eruptions Caused Extinction 250 Million Years Ago." Press release, October 2, 2017. https://www.nyu.edu/about/news-publications/news/2017/october/scientists-find-evidence-that-siberian-volcanic-eruptions-caused.html.

Nicholas, Elizabeth. "An Incredible Number of Americans Have Never Left Their Home State." Culture Trip, January 19, 2018. https://theculturetrip.com/north-america/usa/articles/an-incredible-number-of-americans-have-never-left-their-home-state/.

Norton, Mary Beth, David M. Katzman, David W. Blight, Howard Chudacoff, and Fredrik Logevall. *A People and a Nation: A History of the United States, Volume 2; Since 1865.* 7th ed. Boston: Wadsworth, 2006.

Nossiter, Adam. "France Suspends Fuel Tax Increase That Spurred Violent Protests." *New York Times*, December 4, 2018. https://www.nytimes.com/2018/12/04/world/europe/france-fuel-tax-yellow-vests.html.

Nuccitelli, Dana. "Is the Climate Consensus 97%, 99.9%, or Is Plate Tectonics a

Hoax?" *Guardian*, May 3, 2017. https://www.theguardian.com/environment/climate-consensus-97-per-cent/2017/may/03/is-the-climate-consensus-97-999-or-is-plate-tectonics-a-hoax.

O'Connor, Flannery. *The Complete Stories*. New York: Farrar, Straus and Giroux, 1971.

Office of the Press Secretary. "President Bush Discusses Global Climate Change." White House, June 11, 2001. https://georgewbush-whitehouse.archives.gov/news/releases/2001/06/20010611-2.html.

———. "President's Statement on Climate Change." White House, July 13, 2001. https://georgewbush-whitehouse.archives.gov/news/releases/2001/07/20010713-2.html.

Ossian, Lisa L. *The Forgotten Generation: American Children and World War II*. Columbia, MO, and London: University of Missouri Press, 2011.

Our Children's Trust. "Juliana v. U.S.-Climate Lawsuit." Accessed January 24, 2019. https://www.ourchildrenstrust.org/us/federal-lawsuit/.

Oxfam. "Extreme Carbon Inequality." Oxfam media briefing, December 2, 2015. https://www.oxfam.org/sites/www.oxfam.org/files/file_attachments/mb-extreme-carbon-inequality-021215-en.pdf.

Paglen, Trevor. *The Last Pictures Project*. Video. Creative Time. YouTube, March 20, 2017. https://www.youtube.com/watch?v=dsBJTBKQh9I.

Parisienne, Theodore, Thomas Tracy, Adam Shrier, and Larry McShane. "Famed Gay Rights Lawyer Sets Himself on Fire at Prospect Park in Protest Suicide Against Fossil Fuels." *Daily News*, April 14, 2018. https://www.nydailynews.com/new-york/charred-body-found-prospect-park-walking-path-article-1.3933598.

Parker, Laura. "143 Million People May Soon Become Climate Migrants." *National Geographic*, March 19, 2018. https://news.nationalgeographic.com/2018/03/climate-migrants-report-world-bank-spd/.

Pasiakos, Stefan M., Sanjiv Agarwal, Harris R. Lieberman, and Victor L. Fulgoni III. "Sources and Amounts of Animal, Dairy, and Plant Protein Intake of US Adults in 2007–2010." *Nutrients* 7, no. 8 (2015): 7058–69. https://doi.org/10.3390/nu7085322.

PBS. "Civil Rights: Japanese Americans—Minorities." Accessed February 1, 2019. https://www.pbs.org/thewar/at_home_civil_rights_minorities.htm.

Penn, Justin L., Curtis Deutsch, Jonathan L. Payne, and Erik A. Sperling. "Temperature-Dependent Hypoxia Explains Biogeography and Severity of End-Permian Marine Mass Extinction." *Science* 362, no. 6419 (December 2018). https://doi.org/10.1126/science.aat1327.

Pensoft Publishers. "Bees, Fruits and Money: Decline of Pollinators Will Have Severe Impact on Nature and Humankind." *ScienceDaily*. Accessed March 15, 2019. www.sciencedaily.com/releases/2012/09/120904101128.htm.

Perrin, Andrew. "Who Doesn't Read Books in America?" Pew Research Center,

March 23, 2018. www.pewresearch.org/fact-tank/2018/03/23/who-doesnt-read-books-in-america/.

Perrone, Catherine, and Lauren Handley. "Home Front Friday: The Victory Speed Limit." National WWII Museum. Accessed January 12, 2019. http://nww2m.com/2015/12/home-front-friday-get-in-the-scrap/.

Physikalisch-Meteorologische Observatorium Davos / World Radiation Center (PMOD/WRC). "Solar Constant: Construction of a Composite Total Solar Irradiance (TSI) Time Series from 1978 to the Present." https://www.pmodwrc.ch/en/research-development/solar-physics/tsi-composite/.

Pierre-Louis, Kendra. "Ocean Warming Is Accelerating Faster Than Thought, New Research Finds." *New York Times*, January 10, 2019. https://www.nytimes.com/2019/01/10/climate/ocean-warming-climate-change.html.

Pilon, Mary. "I Found a Dead Body on My Morning Run—It's Something You Can't Run Away From." *Runner's World*, April 18, 2018. https://www.runnersworld.com/runners-stories/a19843617/i-found-a-dead-body-on-my-morning-runits-something-you-cant-run-away-from/.

Plastic Pollution Coalition. "The Last Plastic Straw Movement." Accessed January 25, 2019. https://www.plasticpollutioncoalition.org/no-straw-please/.

Plumer, Brad. "U.S. Carbon Emissions Surged in 2018 Even as Coal Plants Closed." *New York Times*, January 8, 2019. https://www.nytimes.com/2019/01/08/climate/greenhouse-gas-emissions-increase.html.

Poirier, Agnès. "One of History's Most Romantic Photographs Was Staged." BBC, February 14, 2017. www.bbc.com/culture/story/20170213-the-iconic-photo-that-symbolises-love.

Power, Samantha. *"A Problem from Hell": America and the Age of Genocide.* New York: HarperCollins, 2003.

Prairie Climate Center. "Four Degrees of Separation: Lessons from the Last Ice Age." October 28, 2016. prairieclimatecentre.ca/2016/10/four-degrees-of-separation-lessons-from-the-last-ice-age/.

Project Drawdown. "Solutions." Accessed February 1, 2019. www.drawdown.org/solutions.

Pursell, Weimer. "When you ride ALONE you ride with Hitler!" 1943. Poster. National Archives and Records Administration. https://www.archives.gov/exhibits/powers_of_persuasion/use_it_up/images_html/ride_with_hitler.html.

Raftery, Adrian E., Alec Zimmer, Dargan M. W. Frierson, Richard Startz, and Peiran Liu. "Less Than 2°C Warming by 2100 Unlikely." *Nature Climate Change* 7 (2017): 637–41. https://www.nature.com/articles/nclimate3352#article-info.

Rasmussen, Frederick N. "Liberty Ships Honored Blacks in U.S. History." *Baltimore Sun*, March 6, 2004. https://www.baltimoresun.com/news/bs-xpm-2004-03-06-0403060173-story.html.

Reaves, Jessica. "Where's the Beef (in the Teenage Diet)?" *Time*, January 30, 2003. content.time.com/time/health/article/0,8599,412343,00.html.

Rebuild by Design. "The Big U." Accessed January 30, 2019. www.rebuildbydesign.org/our-work/all-proposals/winning-projects/big-u.

Reinert, Al. "The Blue Marble Shot: Our First Complete Photograph of Earth." *Atlantic*, April 12, 2011. https://www.theatlantic.com/technology/archive/2011/04/the-blue-marble-shot-our-first-complete-photograph-of-earth/237167/.

Rennell, Tony. "The Blitz 70 Years On: Carnage at the Café de Paris." *Daily Mail*, April 9, 2010. https://www.dailymail.co.uk/femail/article-1264532/The-blitz-70-years-Carnage-Caf-Paris.html.

Revkin, Andrew C. "Global Warming and the 'Tyranny of Boredom.'" *New York Times*, October 27, 2010. https://dotearth.blogs.nytimes.com/2010/10/27/global-warming-and-the-tyranny-of-boredom/.

Rice, Doyle. "Yes, Chicago Will Be Colder Than Antarctica, Alaska and the North Pole on Wednesday." *USA Today*, January 29, 2019. https://www.usatoday.com/story/news/nation/2019/01/29/polar-vortex-2019-chicago-colder-than-antarctica-alaska-north-pole/2715979002/.

Rich, Nathaniel. "Losing Earth: The Decade We Almost Stopped Climate Change." *New York Times Magazine*, August 1, 2018. https://www.nytimes.com/interactive/2018/08/01/magazine/climate-change-losing-earth.html.

Riding, Alan. "In Louvre, New Room with View of 'Mona Lisa.'" *New York Times*, April 6, 2005. https://www.nytimes.com/2005/04/06/arts/design/in-louvre-new-room-with-view-of-mona-lisa.html.

Ritchie, Earl J. "Exactly How Much Has the Earth Warmed? And Does It Matter?" *Forbes*, September 7, 2018. https://www.forbes.com/sites/uhenergy/2018/09/07/exactly-how-much-has-the-earth-warmed-and-does-it-matter/#7d0059185c22.

Ritchie, Hannah. "How Do We Reduce Antibiotic Resistance from Livestock?" Our World in Data, November 16, 2017. https://ourworldindata.org/antibiotic-resistance-from-livestock.

Robinson, Alexander, Reinhard Calov, and Andrey Ganopolski. "Multistability and Critical Thresholds of the Greenland Ice Sheet." *Nature Climate Change* 2 (2012): 429–32. https://doi.org/10.1038/nclimate1449.

Rochat, Philippe. "Five Levels of Self-Awareness as They Unfold Early in Life." *Consciousness and Cognition* 12 (2003): 717–31. http://www.psychology.emory.edu/cognition/rochat/Rochat5levels.pdf.

Roosevelt, Franklin Delano. "Executive Order 9250, Providing for the Stabilizing of the National Economy." October 3, 1942. Accessed February 2, 2019. https://www.archives.gov/federal-register/executive-orders/1942.html.

———. "Fireside Chat 21: On Sacrifice." April 28, 1942. Miller Center, University of Virginia. Accessed January 30, 2019. https://millercenter.org/the-presidency/presidential-speeches/april-28-1942-fireside-chat-21-sacrifice.

Rosener, Ann. *Women in Industry. Gas Mask Production. . . .* July 1942. Photograph. United States Office of War Information, Library of Congress, https://www.loc.gov/item/2017693574/.

Rothman, Lily, and Arpita Aneja. "You Still Don't Know the Whole Rosa Parks Story." *Time*, November 30, 2015. time.com/4125377/rosa-parks-60-years-video/.

Rumble, Taylor-Dior. "Claudette Colvin: The 15-Year-Old Who Came Before Rosa Parks." BBC World Service, March 10, 2018. https://www.bbc.com/news/stories-43171799.

"Russia's Rich Hiring Luxurious 'Ambulance-Taxis' to Beat Moscow's Traffic Jams." *National Post*, March 22, 2013. https://nationalpost.com/news/russias-rich-hiring-luxurious-ambulance-taxis-to-beat-moscows-traffic-jams.

Safire, William. *Before the Fall: An Inside View of the Pre-Watergate White House.* New York: Doubleday, 1975.

Salk Institute for Biological Studies. "About Jonas Salk." Accessed January 30, 2019. https://www.salk.edu/about/history-of-salk/jonas-salk/.

Scheiber, Noam. "Google Workers Reject Silicon Valley Individualism in Walkout." *New York Times*, November 6, 2018. https://www.nytimes.com/2018/11/06/business/google-employee-walkout-labor.html.

Schein, Lisa. "More People Die from Suicide Than from Wars, Natural Disasters Combined." *VOA News*, September 4, 2014. https://www.voanews.com/a/more-people-die-from-suicide-than-from-wars-natural-disasters-combined/2438749.html.

Schiller, Ben. "Sorry, Buying a Prius Won't Help with Climate Change." *Fast Company*, January 31, 2014. https://www.fastcompany.com/3025359/sorry-buying-a-prius-wont-help-with-climate-change.

Schleussner, Carl Friedrich, Tabea K. Lissner, Erich M. Fischer, Jan Wohland, Mahé Perrette, Antonius Golly, Joeri Rogelj, Katelin Childers, Jacob Schewe, Katja Frieler, Matthias Mengel, William Hare, and Michiel Schaeffer. "Differential Climate Impacts for Policy-Relevant Limits to Global Warming: The Case of 1.5°C and 2°C." *Earth System Dynamics* 7, no. 21 (April 2016): 327–51. https://doi.org/10.5194/esd-7-327-2016.

Schwartz, Alexandra. "Esther Perel Lets Us Listen in on Couples' Secrets." *New Yorker*, May 31, 2017. https://www.newyorker.com/culture/cultural-comment/esther-perel-lets-us-listen-in-on-couples-secrets.

Schwartz, Jason. "MSNBC's Surging Ratings Fuel Democratic Optimism." *Politico*, April 11, 2018. https://www.politico.com/story/2018/04/11/msnbc-democrats-ratings-cnn-fox-513388.

Scott, Michon, and Rebecca Lindsey. "What's the Hottest Earth's Ever Been?" *ClimateWatch Magazine*, August 12, 2014. https://www.climate.gov/news-features/climate-qa/whats-hottest-earths-ever-been.

Scranton, Roy. "Learning How to Die in the Anthropocene." *New York Times*, November 10, 2013. https://opinionator.blogs.nytimes.com/2013/11/10/learning-how-to-die-in-the-anthropocene/.

———. "Raising My Child in a Doomed World." *New York Times*, July 16, 2018. https://www.nytimes.com/2018/07/16/opinion/climate-change-parenting.html.

Sentience Institute. "US Factory Farming Estimates (Animals Alive at Present)." Spreadsheet. https://docs.google.com/spreadsheets/d/1iUpRFOPmAE5IO4hO4PyS4MP_kHzkuM_-soqAyVNQcJc/edit#gid=0.

Shah, Bela. "Addicted to Selfies: I Take 200 Snaps a Day." BBC News, February 27, 2018. https://www.bbc.com/news/newsbeat-43197018.

Shampo, Marc A., and Robert A. Kyle. "Jonas E. Salk—Discoverer of a Vaccine Against Poliomyelitis." *Mayo Clinic Proceedings* 73, no. 12 (1998): 1176. https://doi.org/10.4065/73.12.1176.

Shapiro, Robert Moses, and Tadeusz Epsztein, eds. With an introduction by Samuel D. Kassow. "The Warsaw Ghetto Oyneg Shabes—Ringelblum Archive Catalog and Guide." United States Holocaust Memorial Museum. Accessed January 25, 2019. https://www.ushmm.org/research/publications/academic-publications/full-list-of-academic-publications/the-warsaw-ghetto-oyneg-shabesringelblum-archive-catalog-and-guide.

Shaw, Stacy. "The Overview Effect." *Psychology in Action*, January 1, 2017. https://www.psychologyinaction.org/psychology-in-action-1/2017/01/01/the-overview-effect.

Sifferlin, Alexandra. "Global Jewish Population Approaches Pre-Holocaust Levels." *Time*, June 29, 2015. time.com/3939972/global-jewish-population/.

Smithsonian National Air and Space Museum. "Apollo to the Moon." Accessed January 24, 2019. https://airandspace.si.edu/exhibitions/apollo-to-the-moon/online/later-missions/apollo-17.cfm.

Solly, Meilan. "How Did the 'Great Dying' Kill 96 Percent of Earth's Ocean-Dwelling Creatures?" *Smithsonian*, December 11, 2018. https://www.smithsonianmag.com/smart-news/how-did-great-dying-kill-96-percent-earths-ocean-dwelling-creatures-180970992/.

Springmann, Marco, Michael Clark, Daniel Mason-D'Croz, Keith Wiebe, Benjamin Leon Bodirsky, Luis Lassaletta, Wim de Vries, Sonja J. Vermeulen, Mario Herrero, Kimberly M. Carlson, Malin Jonell, Max Troell, Fabrice DeClerck, Line J. Gordon, Rami Zurayk, Peter Scarborough, Mike Rayner, Brent Loken, Jess Fanzo, H. Charles J. Godfray, David Tilman, Johan Rockström, and Walter Willett. "Options for Keeping the Food System Within Environmental Limits." *Nature* 562, no. 7728 (October 2018): 519–25. https://doi.org/10.1038/s41586-018-0594-0.

Steinfeld, Henning, and Pierre Gerber. "Livestock Production and the Global Environment: Consume Less or Produce Better?" *Proceedings of the National Academy of Sciences* 107, no. 43 (October 26, 2010). https://www.ncbi.nlm.nih.gov/pmc/articles/PMC2972985/pdf/pnas.201012541.pdf.

Steinfeld, Henning, Pierre Gerber, Tom Wassenaar, Vincent Castel, Mauricio Rosales, and Cees de Haan. *Livestock's Long Shadow: Environmental Issues and Options*. Rome: Food and Agriculture Organization of the United Nations, 2006. http://www.fao.org/docrep/010/a0701e/a0701e.pdf.

Steinfeld, Henning, and Tom Wassenaar. "The Role of Livestock Production in Carbon and Nitrogen Cycles," *Annual Review of Environment and Resources*, vol. 32 (November 21, 2007): 271–94, https://doi.org/10.1146/annurev.energy.32.041806.143508.

Steinmetz, Katy. "See Obama's 20-Year Evolution on LGBT Rights." *Time*, April 10, 2015. time.com/3816952/obama-gay-lesbian-transgender-lgbt-rights/.

Strain, Daniel. "How Much Carbon Does the Planet's Vegetation Hold?" *Future Earth Blog*, January 31, 2018. www.futureearth.org/blog/2018-jan-31/how-much-carbon-does-planets-vegetation-hold.

Sudhir, K., Subroto Roy, and Mathew Cherian. "Do Sympathy Biases Induce Charitable Giving? The Effects of Advertising Content." Cowles Foundation for Research in Economics, Yale University, November 2015. https://cowles.yale.edu/sites/default/files/files/pub/d19/d1940.pdf.

Sullivan, Patricia. "Bus Ride Shook a Nation's Conscience." *Washington Post*, October 25, 2005. http://www.washingtonpost.com/wp-dyn/content/article/2005/10/24/AR2005102402053.html.

"Super-Sizing the Chicken, 1923–Present." United Poultry Concerns, February 19, 2015. www.upc-online.org/industry/150219_super-sizing_the_chicken.html.

Taagepera, Rein. "Size and Duration of Empires: Growth-Decline Curves, 600 B.C. to 600 A.D." *Social Science History* 3, no. 3–4 (1979): 115–38. https://doi.org/10.2307/1170959.

Thaler, Richard H., and Cass R. Sunstein. "Easy Does It: How to Make Lazy People Do the Right Thing." *New Republic*, April 2008. https://newrepublic.com/article/63355/easy-does-it.

"The Theft That Made the 'Mona Lisa' a Masterpiece." NPR, July 30, 2011. https://www.npr.org/2011/07/30/138800110/the-theft-that-made-the-mona-lisa-a-masterpiece.

Thompson, A. C. "Timeline: The Science and Politics of Global Warming." *Frontline*. PBS. Accessed January 24, 2019. https://www.pbs.org/wgbh/pages/frontline/hotpolitics/etc/cron.html.

Tillman, Barrett. *D-Day Encyclopedia: Everything You Want to Know About the Normandy Invasion.* Washington, D.C.: Regnery History, 2014.

Truth Initiative. "Why Are 72% of Smokers from Lower-Income Communities?" January 24, 2018. https://truthinitiative.org/news/why-are-72-percent-smokers-lower-income-communities.

Union of Concerned Scientists. "How Do We Know That Humans Are the Major Cause of Global Warming?" August 1, 2017. https://www.ucsusa.org/global-warming/science-and-impacts/science/human-contribution-to-gw-faq.html.

United Nations. "Statement by His Excellency Dr. Fakhruddin Ahmed, Honorable Chief Adviser of the Government of the People's Republic of Bangladesh," at the High-Level Event on Climate Change, United Nations, New York, September 24, 2007. http://www.un.org/webcast/climatechange/highlevel/2007/pdfs/bangladesh-eng.pdf.

United Nations Department of Economic and Social Affairs, Population Division. "World Population Prospects: The 2017 Revision." New York: United Nations, 2017.

United States Bureau of Labor Statistics. "Employment Projections Program." Accessed January 30, 2019. https://www.bls.gov/emp/tables/employment-by-major-industry-sector.htm.

United States Bureau of the Census. "Census of Agriculture, 1969 Volume II." Accessed January 30, 2019. http://usda.mannlib.cornell.edu/usda/AgCensusImages/1969/02/03/1969-02-03.pdf.

United States Climate Change Science Program. "The Climate Change Research Initiative." 2003. Accessed January 24, 2019. https://data.globalchange.gov/assets/2a/42/f55760db8a810e1fba12c67654dc/ccsp-strategic-plan-2003.pdf.

United States Department of Agriculture Economic Research Service. "Access to Affordable and Nutritious Food: Measuring and Understanding Food Deserts and Their Consequences." 2009.

United States Department of the Interior. "New Analysis Shows 2018 California Wildfires Emitted as Much Carbon Dioxide as an Entire Year's Worth of Electricity." November 30, 2018. https://www.doi.gov/pressreleases/new-analysis-shows-2018-california-wildfires-emitted-much-carbon-dioxide-entire-years.

United States Elections Project. "2014 November General Election Turnout Rates." Accessed January 30, 2019. www.electproject.org/2014g.

———. "2016 November General Election Turnout Rates." Accessed January 30, 2019. www.electproject.org/2016g.

United States Energy Information Administration. "Chinese Coal-Fired Electricity Generation Expected to Flatten as Mix Shifts to Renewables." September 27, 2017. https://www.eia.gov/todayinenergy/detail.php?id=33092.

United States Environmental Protection Agency. "Climate Change Indicators: Atmospheric Concentrations of Greenhouse Gases." January 23, 2017. https://www.epa.gov/climate-indicators/climate-change-indicators-atmospheric-concentrations-greenhouse-gases.

———. "Earthrise—the Picture That Inspired the Environmental Movement." Science Wednesday, *EPA Blog*, July 1, 2009. https://blog.epa.gov/2009/07/01/science-wednesday-Earthrise/.

———. "Greenhouse Gas Biogenic Sources, 14.4: Enteric Fermentation—Greenhouse Gases, Supplement D." Chap. 44 in *Air Pollutant Emissions Factors*, 5th ed., vol. 1, February 1998. https://www3.epa.gov/ttnchie1/ap42/ch14/final/c14s04.pdf.

———. "International Treaties and Cooperation About the Protection of the

Stratospheric Ozone Layer." Accessed January 24, 2019. https://www.epa.gov/ozone-layer-protection/international-treaties-and-cooperation-about-protection-stratospheric-ozone.

———. *Inventory of U.S. Greenhouse Gas Emissions and Sinks, 1990–2016*. https://www.epa.gov/ghgemissions/inventory-us-greenhouse-gas-emissions-and-sinks.

United States Holocaust Memorial Museum. "Children During the Holocaust." Accessed March 10, 2019. https://encyclopedia.ushmm.org/content/en/article/children-during-the-holocaust.

University of Illinois Extension. "Turkey Facts." Accessed January 30, 2019. https://extension.illinois.edu/turkey/turkey_facts.cfm.

"U.S. Air Passengers' Main Trip Purposes in 2017, by Type." Statista. Accessed January 31, 2019. https://www.statista.com/statistics/539518/us-air-passengers-main-trip-purposes-by-type/.

Vidal, John. "Protect Nature for Worldwide Economic Security, Warns UN Biodiversity Chief." *Guardian*, August 16, 2010. https://www.theguardian.com/environment/2010/aug/16/nature-economic-security.

Virginia Museum of History and Culture. "Turning Point: World War II." Accessed January 24, 2019. https://www.virginiahistory.org/collections-and-resources/virginia-history-explorer/civil-rights-movement-virginia/turning-point.

Wade, Lizzie. "Tesla's Electric Cars Aren't as Green as You Might Think." *Wired*, March 31, 2016. https://www.wired.com/2016/03/teslas-electric-cars-might-not-green-think/.

Wakabayashi, Daisuke, Erin Griffith, Amie Tsang, and Kate Conger. "Google Walkout: Employees Stage Protest Over Handling of Sexual Harassment." *New York Times*, November 1, 2018. https://www.nytimes.com/2018/11/01/technology/google-walkout-sexual-harassment.html?module=inline.

Wallace-Wells, David. "Could One Man Single-Handedly Ruin the Planet?" *New York*, October 31, 2018. nymag.com/intelligencer/2018/10/bolsanaros-amazon-deforestation-accelerates-climate-change.html.

———. "The Uninhabitable Earth, Annotated Edition." *New York*, July 10, 2017. nymag.com/intelligencer/2017/07/climate-change-earth-too-hot-for-humans-annotated.html.

———. *The Uninhabitable Earth: Life After Warming*. New York: Tim Duggan Books, 2019.

Walters, Daniel. "What's Their Beef? More and More Americans Are Becoming Vegetarians." *Transitions*. Accessed February 5, 2019. https://www.whitworth.edu/Alumni/Transitions/Articles/Calling/TheyretheOtherWhiteMeat.htm.

Weisman, Alan. "Earth Without People." *Discover*, February 2005. discovermagazine.com/2005/feb/earth-without-people.

Wilder, Emily. "Bees for Hire: California Almonds Become Migratory Colonies' Biggest Task." *. . . & the West Blog*, Bill Lane Center for the American

West, Stanford University, August 17, 2018. https://west.stanford.edu/news/blogs/and-the-west-blog/2018/bees-for-hire-california-almonds-now-are-migratory-colonies-biggest-task.

Williams, Casey. "These Photos Capture the Startling Effect of Shrinking Bee Populations." *Huffington Post*, April 7, 2016. https://www.huffingtonpost.com/entry/humans-bees-china_us_570404b3e4b083f5c6092ba9.

Wilson, Michael. "His Body Was Behind the Wheel for a Week Before It Was Discovered. This Was His Life." *New York Times*, October 23, 2018. https://www.nytimes.com/2018/10/23/nyregion/man-found-dead-in-car-new-york.html.

Wise, Irvin L., and Lester M. Hall. Distorting contact lenses for animals. U.S. Patent 3,418,978, filed November 30, 1966. https://patents.google.com/patent/US3418978?oq=patent:3418978.

Wise, Jeff. *Extreme Fear: The Science of Your Mind in Danger.* New York: Palgrave Macmillan, 2009.

Wolf, Julia, Ghassem Asrar, and Tristam West. "Revised Methane Emissions Factors and Spatially Distributed Annual Carbon Fluxes for Global Livestock." *Carbon Balance and Management* 12, no. 16 (2017). https://doi.org/10.1186/s13021-017-0084-y.

Worland, Justin. "Climate Change Used to Be a Bipartisan Issue. Here's What Changed." *Time*, July 27, 2017. time.com/4874888/climate-change-politics-history/.

———. "These Cities May Soon Be Uninhabitable Thanks to Climate Change." *Time*, October 26, 2015. time.com/4087092/climate-change-heat-wave/.

World Bank. *Turn Down the Heat: Climate Extremes, Regional Impacts, and the Case for Resilience: A Report for the World Bank by the Potsdam Institute for Climate Impact Research and Climate Analytics.* Washington, D.C.: World Bank, 2013. http://www.worldbank.org/content/dam/Worldbank/document/Full_Report_Vol_2_Turn_Down_The_Heat_%20Climate_Extremes_Regional_Impacts_Case_for_Resilience_Print%20version_FINAL.pdf.

———. "World Bank Open Data." Accessed January 31, 2019. https://data.worldbank.org/country.

World Food Program. "World Hunger Again on the Rise, Driven by Conflict and Climate Change, New UN Report Shows." September 15, 2017. https://www.wfp.org/news/news-release/world-hunger-again-rise-driven-conflict-and-climate-change-new-un-report-says.

World Health Organization. "Climate Change and Human Health—Risks and Responses." 2003. https://www.who.int/globalchange/climate/summary/en/index5.html.

———. "Fact Sheet: Suicide." August 24, 2018. https://www.who.int/news-room/fact-sheets/detail/suicide.

WorldSpaceFlight. "Astronaut/Cosmonaut Statistics." Accessed January 31, 2019. https://www.worldspaceflight.com/bios/stats.php.

World Wildlife Fund. "Forest Conversion." Accessed January 31, 2019. wwf.panda.org/our_work/forests/deforestation_causes/forest_conversion/.

———. "Wildlife in a Warming World: The Effects of Climate Change on Biodiversity." 2018. https://www.worldwildlife.org/publications/wildlife-in-a-warming-world-the-effects-of-climate-change-on-biodiversity.

Wynes, Seth, and Kimberly A. Nicholas. "The Climate Mitigation Gap: Education and Government Recommendations Miss the Most Effective Individual Actions." *Environmental Research Letters* 12 (2017), 074024. http://iopscience.iop.org/article/10.1088/1748-9326/aa7541/pdf.

Xerces Society for Invertebrate Conservation. "Bumblebee Conservation." Accessed January 30, 2019. https://xerces.org/bumblebees/.

Yaden, David B., Jonathan Iwry, Kelley Slack, Johannes C. Eichstaedt, Yukun Zhao, George Vaillant, and Andrew Newberg. "The Overview Effect: Awe and Self-Transcendent Experience in Space Flight." *Psychology of Consciousness: Theory, Research, and Practice* 3, no. 1 (2016): 1–11. https://doi.org/10.1037/cns0000086.

Zhao, Chuang, Bing Liu, Shilong Piao, Xuhui Wang, David B. Lobell, Yao Huang, Mengtian Huang, Yitong Yao, Simona Bassu, Philippe Ciais, Jean-Louis Durand, Joshua Elliott, Frank Ewert, Ivan A. Janssens, Tao Li, Erda Lin, Qiang Liu, Pierre Martre, Christoph Müller, Shushi Peng, Josep Peñuelas, Alex C. Ruane, Daniel Wallach, Tao Wang, Donghai Wu, Zhuo Liu, Yan Zhu, Zaichun Zhu, and Senthold Asseng. "Temperature Increase Reduces Global Yields." *Proceedings of the National Academy of Sciences* 114 no. 35 (August 2017): 9326–31. https://doi.org/10.1073/pnas.1701762114.

Ziegler, Jean. "Burning Food Crops to Produce Biofuels Is a Crime Against Humanity." *Guardian*, November 26, 2013. https://www.theguardian.com/global-development/poverty-matters/2013/nov/26/burning-food-crops-biofuels-crime-humanity.

Zijdeman, Richard, and Filipa Ribeira da Silva. "Life Expectancy at Birth." Clio Infra. Accessed January 30, 2019. http://hdl.handle.net/10622/LKYT53.

Zimmer, Carl. "The Planet Has Seen Sudden Warming Before. It Wiped Out Almost Everything." *New York Times*, December 7, 2018. https://www.nytimes.com/2018/12/07/science/climate-change-mass-extinction.html.

Zug, James. "Stolen: How the Mona Lisa Became the World's Most Famous Painting." *Smithsonian*, June 15, 2011. https://www.smithsonianmag.com/arts-culture/stolen-how-the-mona-lisa-became-the-worlds-most-famous-painting-16406234/.

Zuidhof, M. J., B. L. Schneider, V. L. Carney, D. R. Korver, and F. E. Robinson. "Growth, Efficiency, and Yield of Commercial Broilers from 1957, 1978, and 2005." *Poultry Science* 93 no. 12 (December 2014): 2970–82. https://doi.org/10.3382/ps.2014-04291.

Acknowledgments

This book began with a conversation Ev Williams and I had in 2017. Soon thereafter he introduced me to Abbey Banks. The two have been generous partners throughout this process and helped me to believe that significant change is possible.

Simone Friedman is, as my grandmother would have said, "a force of nature." Her energy, wisdom, ambition, and optimism bring even the most idealistic visions within reach. The first step toward making the necessary changes in our lives is knowing what changes are needed. Because of Simone's work, along with that of Manny Friedman and EJF Philanthropies, the all-important connection between climate change and animal agriculture is finally in the public consciousness.

I hired Tess Gunty as a research assistant, but she quickly became my first reader, and ultimately my collaborator. Every sentence of this book benefited from her thoughtfulness.

I can't think of any subjects that are more complex and controversial than the planetary crisis and our food choices. Hunter Braithwaite's vigilant fact-checking was indispensable.

I was in communication with numerous climate science experts while writing this book. I am thankful for all the time, information, and knowledge they shared. Brent Kim, Raychel Santo, and Jeff Anhang deserve special mention.

Farrar, Straus and Giroux has once again reminded me of how lucky I am to be a writer. I am particularly grateful to Scott Auerbach, Rodrigo Corral, Jonathan Galassi, M. P. Klier, Spenser Lee, Jonathan Lippincott, Alex Merto, June Park, Julia Ringo, and Jeff Seroy.

As much as anything else, this book is about home. Nicole Aragi and Eric Chinski have been my literary home for almost twenty years. Thank you.

A NOTE ABOUT THE AUTHOR

Jonathan Safran Foer is the author of the novels *Everything Is Illuminated*, *Extremely Loud and Incredibly Close*, and *Here I Am*, and of the nonfiction book *Eating Animals*. His work has received numerous awards and has been translated into thirty-six languages. He lives in Brooklyn.